INNOVATION IN AGRI-FOOD CLUSTERS

Theory and Case Studies

Innovation in Agri-food Clusters

Theory and Case Studies

Peter W.B. Phillips, Jeremy Karwandy,
Graeme Webb and Camille D. Ryan

www.cabi.org

CABI is a trading name of CAB International

CABI
Nosworthy Way
Wallingford
Oxfordshire, OX10 8DE
UK

Tel: +44 (0)1491 832111
Fax: +44 (0)1491 833508
E-mail: info@cabi.org
Website: www.cabi.org

CABI
38 Chauncey Street
Suite 1002
Boston, MA 02111
USA

Tel: +1 800 552 3083 (toll free)
Tel: +1 (0)617 395 4051
E-mail: cabi-nao@cabi.org

A catalogue record for this book is available from the British Library, London, UK.

Library of Congress Cataloging-in-Publication Data

Innovation in agri-food clusters : theory and case studies / Peter W.B. Phillips ... [et al.].
p. cm.
Includes bibliographical references and index.
ISBN 978-1-78064-041-9 (hb)
1. Agricultural innovations--Saskatchewan--Saskatoon--Case studies. 2. Food industry and trade--Saskatchewan--Saskatoon--Case studies. 3. Agricultural industries--Saskatchewan--Saskatoon--Case studies. I. Phillips, Peter W. B.
S494.5.I5I5534 2012
338.1097124'25--dc23
2012036038

ISBN: 978 1 78064 041 9

Commissioning editor: Claire Parfitt
Editorial assistant: Chris Shire
Production editor: Tracy Head

Typeset by Columns Design XML Ltd, Reading, UK.
Printed and bound in the UK by CPI Group (UK) Ltd, Croydon, CR0 4YY.

Contents

About the Authors

Dr Peter W.B. Phillips is Professor of Public Policy in the Johnson-Shoyama Graduate School of Public Policy at the University of Saskatchewan, Saskatoon, Canada. He earned his PhD in International Political Economy at the London School of Economics and practised for 13 years as a professional economist and senior policy advisor in Canadian industry and government. At the University of Saskatchewan, he has held the Van Vliet Research Chair, created and held an NSERC-SSHRC Chair in Managing Technological Change, was a founding member and director of the virtual College of Biotechnology and was founding director of the Graduate School of Public Policy. He has had visiting appointments at the LSE, the OECD, the European University Institute and the University of Western Australia, is associate editor of AgBioForum, a leading on-line journal, was a member of the NAFTA Chapter 13 expert panel on GM maize in Mexico and was a founding member of the Canadian Biotechnology Advisory Committee. He has been a member of the Board of Directors of the Canadian Agri-food Policy Institute, the Estey Centre for the Study of Trade, Law and the Economy, and Ag-West Bio Inc., which operates a biotech venture fund. His current research focuses on governing transformative innovation, including regulation and policy, innovation systems, intellectual property management, trade policy and decision systems. He is co-lead and principal investigator of a CAN$5.4 million Genome Canada project entitled Value Addition through Genomics and GE3LS (VALGEN), which runs from 2009–2013, and has been an applicant and investigator on more than 15 peer-reviewed grants worth more than CAN$150 million. He has been author or editor of eight other books and more than 70 journal articles and book chapters.

Jeremy Karwandy is an economic development practitioner and policy analyst with Enterprise Saskatchewan in Saskatoon, Canada. He earned his MSc through the interdisciplinary graduate programme at the University of Saskatchewan by investigating the interface between theories of the firm and theories of economic agglomeration. His current work focuses on advancing policies and programmes that enhance the provincial environment for entrepreneurship and overall corporate competitiveness. This includes projects as varied as mentorship programming, entrepreneurship education services, angel investment incentives and productivity measurement.

Graeme Webb is in the PhD program in the School of Communication at Simon Fraser University in Vancouver, Canada. He earned his MA in Political Science at the University of Saskatchewan where he examined the relationship between creative social entrepreneurs, social capital and networked urban governance. From 2007 to 2011, he was actively involved with the Innovation Systems Research Network examining the social dynamics of economic development. He has also been a researcher for the United Nations University Institute on Comparative Regional Integration. His doctoral research combines communication theory and political sociology.

Dr Camille D. Ryan is currently a Professional Research Associate with the Departments of Plant Science and Bioresource Policy, Business and Economics at the University of Saskatchewan, Canada, splitting time between there and her home in the Alberta Foothills. Prior to returning to the University of Saskatchewan to complete her doctoral studies, she worked for both a small plant biotech firm in Saskatoon and in administration with AgrEvo (now Bayer) and was part of the latter firm's efforts to bring Liberty Link (GM) canola to the market in the mid-1990s. She continues to network with the broader science community working on projects with Canadian organizations such as the National Research Council's Industrial Research Assistance Program (IRAP), the International Institute for Sustainable Development, Alberta Advanced Education and Technology, Agri-Food Canada (AAFC) and the Saskatchewan Flax Development Commission. Most recently, she is connected to the Institute of Agriculture at the University of Western Australia working on a project evaluating public–private partnerships in pulse breeding and research in Canada and Australia. She has presented at conferences around the world on various topics including: collaborations in plant genetic resources, and access and benefits sharing, intellectual property and strategic and issues management in agriculture, as well as science communications. She is an avid blogger and Twitterer and is the author of the book *Evaluating Performance of Research Networks: a Socioeconomic Framework for Assessing Funded Research Projects*, a reference and resource for policy makers and project managers.

Preface

This monograph is the result of 10 years of on-going research undertaken as part of the Innovation Systems Research Network (ISRN), a major collaborative research initiative funded by the Social Science and Humanities Research Council of Canada.

In one way, our experience with this work reifies the serendipitous nature of innovation. Each of us engaged in this research more by chance than by design, and our success was due to open collaboration.

Phillips was the Van Vliet Chair Professor in Agricultural Economics at the University of Saskatchewan in 1999 when Marlene Moleski, a colleague with the Canadian Department of Western Diversification, passed his name along to Adam Holbrook, Associate Director of CPROST at Simon Fraser University and an instigator of INNOCOM, a precursor research network to the ISRN. Phillips then made a presentation at a Policy Research Initiative conference in Ottawa; David Wolfe and Meric Gertler, the Principal Investigators on ISRN, were in the audience. Based on those meetings, Phillips was enrolled as a collaborator in the project and has stayed with it as he moved from Agricultural Economics through Political Studies to the Johnson-Shoyama Graduate School of Public Policy at the University of Saskatchewan.

Ryan was pursuing a BComm in the virtual College of Biotechnology in Saskatoon, after a period working in the local biotechnology community. She took a class from Phillips, got a summer job doing some of the early surveying on the project, did her PhD in interdisciplinary studies under Phillips and on the project and is now a professional research scientist in Calgary.

When Webb started graduate studies in the Department of Political Studies at the University of Saskatchewan, he was somewhat arbitrarily assigned to Phillips to supervise. He ended up doing his MA on work relevant to and funded by the ISRN grant and used contacts from the ISRN to arrange doctoral studies at Simon Fraser University.

Karwandy was in the same BComm cohort as Ryan. After a few years in business, he returned to do his MSc in Interdisciplinary Studies under Phillips and on work related to the ISRN grant.

While the ISRN grant gave us the space and resources to undertake this research, the scale, scope and direction of the work was not planned. We individually and collectively followed our muses, ultimately undertaking what we believe is a novel study in terms of its scale and scope of innovation in the agri-food industry.

The one ideal that binds us to the topic, above all else, is that it is both a high calling and moral imperative to efficiently and effectively advance our global capacity to produce enough nutritious food at appropriate prices to feed a burgeoning world population. Without that, nothing else will matter in the end.

Acknowledgements

We gratefully acknowledge the generous support of SSHRC, our collaborators in the ISRN and all of our classmates and colleagues who have contributed to the ideas in this book.

PWBP, JK, GW and CDR
Saskatoon, Vancouver and Calgary

List of Tables

List of Figures

Acronyms and Abbreviations

AAFC	Agriculture and Agri-Food Canada
AB	Alberta
ACOA	Atlantic Canada Opportunities Agency
AS	abiotic stress
BC	betweenness centrality
BC	British Columbia
BERD	business expenditure on research and development
BG	Brassica Genome
CC	closeness centrality
CCC	Canola Council of Canada
CDC-SPG	Crop Development Centre-Saskatchewan Pulse Growers
CFI	Canada Foundation for Innovation
CGIAR	Consultative Group on International Agricultural Research
CIBC	Canadian Imperial Bank of Commerce
CIHR	Canadian Institutes of Health Research
CIRCLE	Centre for Innovation, Research and Competence in the Learning Economy
CIS	community innovation surveys
CLS	Canadian Light Source
CMA	census metropolitan area
CRADA	Collaborative Research and Development Agreement
DUI	doing, using, interacting
EC	Eigenvector centrality
EU	European Union
FPMI	Functional Pathogenomics of Mucosal Immunity
GC	Genome Canada

GDP	gross domestic product
GE^3LS	Genomics, Ethics, Environment, Economics, Law and Society
GERD	gross expenditure on research and development
GLIP	Grain Legumes Integrated Project
GM	genetically modified
HHI	Herfindahl–Hirschman Index
HQP	high quality personnel
ICAP	Institute for Competitiveness and Prosperity
ICARDA	International Center for Agricultural Research in the Dry Areas
ICRISAT	International Crops Research Institute for the Semi-Arid Tropics
ICT	information and communications technology
IRAP	Industrial Research Assistance Program
ISRN	Innovation System Research Network
LQ	location quotient
MAR	Marshall–Arrow–Romer
MNE	multi-national enterprise
NAICS	North American Industry Classification System
NESTI	National Experts on Science and Technology Indicators
NGO	non-governmental organization
NIS	national innovation system
NRC	National Research Council
NRC-IRAP	National Research Council – Industrial Research Assistance Program
NRC-PBI	National Research Council – Plant Biotechnology Institute
NSERC	National Sciences and Engineering Research Council
NZ	New Zealand
OECD	Organisation for Economic Co-operation and Development
ORA	Organizational Risk Analyzer
P3	public–private partnership
PBI	Plant Biotechnology Institute
PBV	product-based view
PRL	Prairie Research Laboratory
R&D	research and development
RBV	resource-based view
RIS	regional innovation system
SD	standard deviation
SIIF	Science and Innovation Investment Framework
SK	Saskatchewan
SNA	social network analysis
SRC	Saskatchewan Research Council
SREDA	Saskatchewan Regional Economic Development Authority
SSHRC	Social Sciences and Humanities Research Council
STI	science, technology and innovation
TDC	total degree centrality
vCB	virtual College of Biotechnology

VIDO	Vaccine and Infectious Disease Organization
WED	Western Economic Diversification
WDR	World Development Report

1

The Blind Leading the Blind along the Innovation Pathway

It was six men of Indostan
To learning much inclined,
Who went to see the Elephant
(Though all of them were blind),
That each by observation
Might satisfy his mind.

The First approach'd the Elephant,
And happening to fall
Against his broad and sturdy side,
At once began to bawl:
'God bless me! but the Elephant
Is very like a wall!'

The Second, feeling of the tusk,
Cried, – 'Ho! what have we here
So very round and smooth and sharp?
To me 'tis mighty clear
This wonder of an Elephant
Is very like a spear!'

The Third approached the animal,
And happening to take
The squirming trunk within his hands,
Thus boldly up and spake:
'I see,' quoth he, 'the Elephant
Is very like a snake!'

The Fourth reached out his eager hand,
And felt about the knee.
'What most this wondrous beast is like
Is mighty plain,' quoth he,
''Tis clear enough the Elephant
Is very like a tree!'

The Fifth, who chanced to touch the ear,
Said: 'E'en the blindest man
Can tell what this resembles most;
Deny the fact who can,
This marvel of an Elephant
Is very like a fan!'

The Sixth no sooner had begun
About the beast to grope,
Then, seizing on the swinging tail
That fell within his scope,
'I see,' quoth he, 'the Elephant
Is very like a rope!'

And so these men of Indostan
Disputed loud and long,
Each in his own opinion
Exceeding stiff and strong,
Though each was partly in the right,
And all were in the wrong!

John Godfrey Saxe

While innovation is the buzz word of the 21st century, it suffers from the parable of the Blind Men and the Elephant. Many bright individuals have studied and practised innovation, leaving us with a loud and long dispute in which each is partly right yet all are wrong. Sifting through these differences in opinion and sorting out the contributions of value is a necessary step towards unlocking the potential for innovation to drive growth and prosperity.

Innovation is almost unanimously recognized as the key factor in the ability of local and national economies to prosper successfully and continuously. There is no sector in which sustained innovation is more important than in the agri-food production sector, which is tasked with feeding a rapidly growing world population. The greatest challenge to the long-term viability and sustainability of the global agri-food sector is its ability to meet or exceed the productivity gains of other sectors in the global economy. If producers and related industrial actors are unable to sustain their comparative productivity, then land, labour and capital will be drawn away from agriculture into higher-performing activities. Thus, the prime goal of any long-term agri-food policy, in virtually any economy that relies on the market to govern production and consumption, must be to assist and enable the sector to innovate and to increase the value added for the factors of production.

Innovation is generally viewed as the adoption of a new, useful or significantly improved idea, good, service, process or practice. University of Minnesota economist Vernon Ruttan (2003) suggested that invention requires generating 'new things' through an 'act of insight', which is consistent with the patent system, which defines invention as something that is new, useful and non-obvious. But innovation involves far more than invention. Ultimately, innovation involves a change in the production system itself. Austrian economist Joseph Schumpeter (1934) concluded that 'we simply define innovation as the setting up of a new production function and "technical change" is a shorthand expression for any kind of shift in the production function'. The by-product of this is social change. As Gladwell (2000, p. 62) suggests, 'innovation – the heart of the knowledge economy – is fundamentally social'.

Recent literature offers mixed advice on how to encourage and sustain innovation. Neoclassical economists argue that, as long as the underlying fundamentals in the economy (inflation, interest rates, exchange rates, wage rates and relative prices) are appropriate, individual consumers and profit-optimizing actors will bring forward the optimal amount of research, development and commercialization of new ideas that can change the production function. A range of regional economists, business strategists and systems theorists posit that, while the fundamentals may be necessary for innovation, they are not sufficient. Rather, they argue that innovation occurs in social systems that are embedded in specific places, in specific processes or in specific groups of creative people.

The parable of the Blind Men and the Elephant, attributed to the Pali Buddhist Udana, compiled in the 2nd century BCE and modernized by John Godfrey Saxe in 1873, is as good a metaphor as any to frame analysis of the multiple facets attributed to regional economic innovation.

A number of aspects of the parable fit nicely with the current thinking about innovation. In the first instance, most of the theorists, thinkers and advisors on innovation come at the topic with some structural 'blindness'. Some bring principles-based theories to the problem, undertaking a deductive exercise of hypothesizing and testing the comparative statics of economic change. Others are more inductive, either bringing observation from hands-on experience or direct investigation of the places, processes and people involved in innovation not just to explain the scale and scope of change, but to investigate one or more of the potential forces that directed that change. All, ultimately, come with blinders or partial blindness to counterfactual explanations.

One can see at least four subsets of investigations that reflect the approaches of the blind men of Indostan. The blind man who bumped into the wall exhibited the general ethos in the economics profession, where they assume an impervious wall around the innovation complex. Instead of examining the functioning of the system, they theorize and hypothesize a range of structures, functions and behaviours, which are then tested to determine how well they explain the external effects of innovation. The two blind men who felt and identified physical structures – the spear and rope – exhibit many of the attributes of those who see clusters as manifesting themselves in agglomerations of fixed capital investment and institutions. The blind man who identified the ear as a fan was thinking intuitively about processes, whereby actions and reactions lead to external effects; many systems theorists similarly see innovation as a nested set of processes that link place and people. Finally, the two men who identified living, organic elements – the snake and tree – were drawn implicitly to the organic entities that make up the system. So, too, many innovation theorists posit that one needs to look at the people in the system to understand the sum of the parts. Moreover, the disharmony – disputing loud and long with opinions stiff and strong – reflects the current debate. Finally, as with all complex problems, it is also likely that 'each was partly in the right, and all were in the wrong!'

The core purpose of this book is to examine agri-food innovation, looking carefully at the theories, assumptions, methods and evidence for each of the four competing paradigms of innovation theory and practice.

THE SCOPE OF INNOVATION POLICY

Governments, industry, scholars and pundits around the world are increasingly focused on innovation as both a means and an end to improving economic and social well-being. Despite the wide range of views about the concept, Michael Porter's version of clusters (1990) has been rapidly adopted by practitioners from all over the world and has become a baseline for much of the debate. Economic development agencies in most developed and developing countries have applied Porter's generalized approach to clusters, customizing it to their particular geopolitical opportunities.

A large number of clusters have been examined in recent years (KPMG-DRI McGraw-Hill-IMPAX Policy Services, 1996; Nordicity Group, 1996; ACOA, 1997; Gollub, 1997; Roelandt and den Hertog, 1998; Scottish Enterprise Network, 2000; van der Linde, 2002; Sölvell *et al.*, 2003; National Research Council Canada, 2004, 2005; Wolfe and Lucas, 2004), but few of these cluster studies were targeted on the agri-food sector. Even though the agri-food sector arguably represents the single largest global economic activity, only about 100 of the approximately 1080 clusters examined in the noted studies examined agri-food activities. Lagnevik *et al.* (2003) provided an analysis of the processes of innovation found in industrial clusters focusing, in particular, on the characteristics of innovation clusters and their operation in the food industry. Furthermore, many of those that were studied tended to be interconnected with other industrial activities, such as wineries and speciality food preparation services that were usually examined as part of a larger tourism sector.

This book aims to address the incomplete study of innovation in this critical global sector by examining the theory and practice in the Saskatoon agri-food sector.

INNOVATION IN THEORY AND PRACTICE

Markusen (1999) asserted that the innovation literature is rife with 'fuzzy concepts' with 'scanty evidence' that lead to 'wimpy policies'. The difficulty is that four competing paradigms that attempt to explain the incidence of innovation have overlapping and interlocking assumptions, hypotheses and methods, yet offer conflicting and incommensurate policy recommendations.

As with much of the literature about technology and the economy, Joseph Schumpeter is credited with the first economic definition of innovation. He defined innovation as the introduction of a new good or a new quality of an existing good, a new method of production, opening of a new market, introduction of a new supply of inputs to a production system or a new organizational structure in an industry (Schumpeter, 1939).

Two points are clear from his definition. First, he very explicitly separates invention from innovation, arguing that 'innovation is possible without anything we should identify as invention, and invention does not necessarily induce innovation' (Schumpeter, 1939). While some might argue that innovation involves introduction of a 'better way of doing things', one should keep in mind that many, if not most, new technologies typically emerge in a relatively primitive condition. It is increasingly accepted that it may be irrelevant whether an innovation offers any advantages over the product, technology, market or organization it is replacing; rather, all that matters is whether individuals perceive an advantage. On that note, discourse and sociopolitical context that surrounds particular technological innovation can drive adoption and evolution of its use. Sustainable innovation, however, most certainly requires a net gain to society. Regardless of these nuances, much of the literature and all of the public policy interest tends to focus on those innovations that bring something new – but especially an invention – into widespread, practical use.

Secondly, Schumpeter is clearly talking about a process ('introduction') by which something new is initiated or adopted. Ruttan (2003) concurred, suggesting that the concept of innovation be extended to include the process by which new things emerge in science, technology and art. A variety of other researchers see creativity – defined as the production of novel and useful ideas in any domain – as the basis for innovation, and so they define innovation as the successful implementation of creative ideas within an organizational structure.

Much of what people identify as innovation involves planned, systematic effort to add value through research and development (R&D). Stanley Metcalfe (1995), professor of political economy and Director of Policy Research in Engineering, Science and Technology at Manchester University, suggested that a number of key stages occur with most innovations: investments are driven by innovation possibilities anticipated in the minds of technologists, research managers and innovators; it is uncertain whether the research, development and commercialization inputs will generate any innovative output; and innovation co-produces targeted useable output (which might be agenda reducing) as well as enhancing the number of possibilities (which is agenda enhancing). The Committee for Study of Invention (2004) produced a report for the US National Science Foundation that explicitly differentiates invention ('the process of devising and producing by independent investigation, experimentation, and mental activity something that is useful and that was not previously known or existing') from innovation, which they define as 'the complex process of introducing novel ideas into use or practice. Inventions, thus, generate value from their use (innovation) and not from their creation and existence.' In essence, innovation and technological change do not follow a pre-determined logic of their own. A series of social, political and economic choices, often manifested in public policy and institutional and organizational effort, drives innovation – technology and the benefits that we derive from technology are then the by-products of this effort.

There are two main competing views about the processes of innovation. The economic theory of technological change focuses on the firm as the primary research unit (e.g. Solow, 1956; Arrow, 1962). Economists then examine the microeconomic incentives for and impacts of private research and commercialization. Some economists recently have examined the impact that innovative activity at the firm level has on the larger economy, focusing especially on the implications of endogenously generated innovation on macroeconomic growth, trade and industrial location. Even so, they tend to assume that innovation is a discrete event that occurs within firms. Recently, a few economists have begun to note that firms are not self-sufficient and often must reach beyond their boundaries to supplement their capacity. Nevertheless, their focus remains steadfastly on innovation that is orchestrated by a single corporate leader or a small subset of purposeful firms.

The expansion of the literature beyond the reductionist neoclassical economic approach and search for deeper meaning was at least partly triggered by the advent of the *new* economics of growth (Lucas, 1988; Romer, 1990), trade (Krugman, 1996) and geography (Krugman, 1998a). While the new economic theories showed that endogenous growth could be exploited to generate economic advantage, they did not explicitly examine the growth processes.

Beginning in the 1990s, political economists, business theorists and sociologists with more of an interest in the influence of institutions, ideas and individuals began to develop a range of theoretical systems approaches to innovation. Most hark back to Alfred Marshall and his *Principles of Economics* (1890) and look at the role of economies of scale and scope in the local, regional, national or sectoral innovation systems themselves.

Now we have a range of discernible, competing theories of innovation. These perspectives can be parsed into four broad, competing theories of R&D, innovation and growth that offer a range of prospective policy advice (Table 1.1). While innovation systems are all the rage in policy circles, growth theorists and development economists abstract from the focus on innovation processes, asserting that optimal investment will be derived in well-structured economies, while a mix of business economists, economic geographers and sociologists posit that innovation is the result of some combination of *special people* doing *special things* in *special places* (Leadbeater, 2005).

While the four interrelated theories of research and development, innovation and growth each make glancing reference to place, processes and people, they tend to focus on one driver, holding the others constant. The recurrence of these three elements across all four approaches adds to the fuzziness of the literature but more importantly underscores the importance of all three to development, innovation and growth.

The earliest and most popular among policy makers is that the pursuit of policies that create special places will generate all these positive outcomes. Marshall (1890) and Porter (1990) both noticed the presence of agglomerations of firms and industries and attempted to explain them in the context of what are now called economies of scale. Firms co-locating in regions are likely to gain through competition and traded

Table 1.1. Four models of innovation.

Assumptions	Neo-classical trade theory	Clusters	Innovation systems	Economies of creativity
Place	Heckscher–Ohlin theorem posits that production locates where comparative advantages match comparative endowments	Agglomeration of firms and industries create economies of scale where inter-dependences are traded (e.g. thick labour market and dense forward and backward linkages)	Untraded inter-dependencies ('mysteries in the air') are geographically sticky	Creatives live in cities endowed by technology, talent and tolerance
Processes	Relative prices (e.g. wages, exchange rates, inflation, interest rates) are more important than organizational structures	Mode 1 system, where MNEs and universities interact in hierarchical processes; both are anchors for clusters	Social networks drive innovation in a mixture of mode 1 and mode 2 systems (e.g. triple helix and regional innovation systems)	Creative people live in purpose-built, heterogeneous, mode 2-style networks
People	Individual optimization drives innovation	Angels and venture capitalists generate demand pull system	Research stars at core of science and technological push system	Creative, talented people (creatives) driven both by personal and commercial motives

interdependencies, such as extensive, supportive forward and backward linkages and thick labour markets (i.e. where both demand for, and supply of, skilled labour is high). Policies directed at nurturing special places tend to encourage firms, sectors or industries to co-locate. Specific efforts include subsides, tax credits and a range of special infrastructures to support firm attraction and operations (e.g. universities, public laboratories and industrial parks). Firms pursuing a product-based view (PBV) of the world, where competition through cost minimization is critical, are most attracted to such centres as they facilitate cost efficiencies that allow them to gain or sustain their targeted market share. Ironically, while the cluster model has been subsumed in the larger innovation literature, there is little explicit focus on the theory or the policy prescriptions on innovation or knowledge management as the main or even an important concern. Efficiency trumps all other motives.

In the 1980s, the special processes approach to innovation gained ascendancy. Variously described as national systems of innovation (Lundvall, 1992; Nelson, 1993; Freeman, 1995), regional systems of innovation (Cooke, 1998; Wolfe and Gertler, 2004) and the triple-helix model (Etzkowitz and Leydesdorff, 1995, 2000), the special processes approach posits that economies of scope generate untraded interdependencies between people and firms that coexist in innovation systems. The frequent and complex interactions of individuals and firms as they develop and test new ideas in the market create social capital that facilitates the transmission of information and knowledge. Bourdieu (2006, p. 98) explored this to a greater or lesser degree but pointed to the 'arbitrariness of appropriation' that 'arises most sharply in the process of transmission' of capital.

Gibbons *et al.* (1994) posited that two modes of knowledge generation now exist: mode 1, which they called traditional knowledge, is generated within disciplinary, primarily cognitive, contexts, while mode 2 knowledge is created in broader, transdisciplinary social and economic contexts. Malecki (1997) parsed the knowledge generated in these two systems into four types, based on their degree of form (codified or tacit) and structure (whether residing in individuals or the collective). He came up with four discernible categories of knowledge: know-why knowledge, such as the laws of nature and heredity, is codified and collective; know-what knowledge, involving recipes for doing things, is frequently codified (in patents) but individual; know-how, the skill of integrating knowledge and dexterity, is tacit and individual; and know-who, knowing who knows what and who knows how to do things, is inherently tacit and collective. This level of detail is needed, because much of what happens in special processes involves the creation and exchange of information and knowledge. Unlike the special places theory, however, these interactions most often happen outside formal transactions – they emerge organically and are a collective product of local culture and context. Policy focused on generating innovation systems tends to be largely supply–push, involving investment in institutions, programmes or processes where research, development and commercialization activities can prosper. Firms attracted to this type of system are likely to be pursuing resource-based strategies (Penrose, 1959), where they focus more on core competencies than on products and market share.

At about the turn of the millennium, a group of business theorists and economic geographers broke away from the special processes group and began to explore the role of the individual in the context of the innovation process (e.g. Florida, 2002a,b; Gertler *et al.*, 2002). A set of interrelated theories and policy prescriptions assume that creative individuals are at the core of the innovation process. A mix of sociology, psychology and economic theories assert that highly creative individuals can be nurtured and unleashed in communities that invest in technology and talent and celebrate tolerance to diversity and change (i.e. Florida's three Ts). The institutional and social pathways of change are not fully delimited in this literature, but the central tendency of those promoting this approach is to generate high-quality, creative human capital through education and investments in research and networking programmes and to invest in talent attraction through nurturing the local quality of life. Firms that are attracted by this type of milieu tend to be pursuing knowledge-based strategies (Grant, 2002), seeking to be on the leading edge of technology, product and organizational innovation. The open-source, creative commons approach is often wistfully cited as the perfect environment for creative types, as it least constrains novel recombinations of knowledge and ideas.

While the separation of these perspectives is an important step towards conceptual clarity, it is important to note that, in practice, none of these perspectives can realistically be separated from the others. Every location will find some form of formal business-to-business linkages alongside informal links to local public or common pool goods, as well as some higher form of motivation (higher than economic motives) to push the local community towards expanding the production frontier. Nevertheless, making the distinction between the different perspectives and their underlying relational mechanisms offers the opportunity to investigate the causal mechanisms and behavioural models so as to test our understanding of when, where and to whom different policy measures are relevant.

In essence, theorists and policy makers in the modern, knowledge-based economy are currently trying to balance competing views of how society can optimize the creation and use of new knowledge. This task has engaged scholars, scientists, policy makers, inventors, entrepreneurs, capitalists, consultants and non-governmental organizations (NGOs) in a heated debate (with both positivist and normative elements) about which model is best.

AGRI-FOOD INNOVATION CASE WORK

One weaknesses of the innovation literature is that it has focused mostly on the newest and sexiest agglomerations of innovative industrial activity in advanced industrial economies, including biomedical, information and communications technology (ICT), advanced instrumentation, high fashion and entertainment. Significant effort has been directed to understanding how innovation happens in largely urban settings, such as Silicon Valley in California and Route 128 in Massachusetts, around top-ranked global universities in major global cities, in the national systems of innovation anchored by global multinational enterprises (MNEs) in the Scandinavian countries, in and around the *keiretsu* in Japan and *chaebol* in South Korea, within fashion-based clusters in Northern Italy, New York and London, and in creative media centres such as Los Angeles, New York and London. The common link in these studies is that they focus largely on exceptional cases – they are the spiky places in the world where exciting and photogenic world leaders are visibly present. The challenge is that most of the world's population and most economic sectors do not exhibit the same levels of exceptionalism. Rather, most sectors and communities toil in relative obscurity. An examination of the exceptions will not generate the rule; a broadly applicable, functioning theory of innovation requires us to turn our attention to the toiling centres of the world.

Saskatoon offers a case study of one such toiling centre that has had some sustained success in the agri-food sector. The period under review in this book starts back in the early days of settlement, but focuses primarily on the emergence of an innovation culture and track record in the past generation.

Cities like Saskatoon, in Canada and around the world, are often overlooked in the context of development. As the most northerly metropolitan city in Canada that is not also a provincial or territorial capital, Saskatoon has spent more than a century attempting to craft an independent identity that reflects the reality of local strengths and resonates beyond the city limits. The city has been buffeted by a series of booms and busts. Global economic cycles in the 20th century at times delivered price surges for globally traded commodities produced nearby and serviced by Saskatoon, which boosted employment, incomes and wealth; all too often these booms were quickly followed by commodity price collapses, which undercut incomes, wiped out wealth and forced released workers to migrate (mostly to Calgary and beyond).

Since the Second World War, Saskatoon by choice and by chance has laid the foundation for what currently appears to be an exciting present and prosperous future. Resource development in Saskatchewan accelerated after the war, differentially helping Saskatoon, as potash was centred on the city and most of the minerals and forestry in the north are served by Saskatoon. Similarly, fundamental economic restructuring in the agri-food sector helped Saskatoon as new crops and livestock opportunities emerged in the dark brown soil zone surrounding the city and across the northern half of the farm belt, which sustained on-farm activity and generated new opportunities for processing or serving the sector.

The Saskatoon economy and population began to accelerate beginning in the 1980s, in the first instance because of rural exodus and later due to consolidations and expansions in the resource area. For much of this period, no one would have said that Saskatoon was an innovation hub. Rather, it was a resource town, with all of the attendant challenges. The single most important shift was the decision to focus on agri-food and biotechnology research at the University of Saskatchewan in Saskatoon. This ultimately put the city on the map as one of the earliest and most successful agricultural biotechnology innovation clusters in the world. The lessons learned in biotechnology – that community leadership, strategic investment in university-based and public research infrastructure and selected industrial support could build a world-class sector – have now been translated into efforts to expand research and development in other resource areas.

As recently as 2000, per capita incomes in Saskatoon lagged behind national averages, relatively slow employment and demographic growth was more related to intra-provincial restructuring than net provincial economic growth, and the asset value of homes, businesses and property was significantly lower than in comparably sized communities elsewhere in Canada. However, with relatively strong global commodity markets in 2000–2010, the situation changed. New and potentially more profitable resource deposits have been delineated and prepared for development, including oil, heavy oil, potash, uranium, gold and diamonds. An array of technologies has also made some of the province's products more profitable. Meanwhile, over the decade, more than CAN$1 billion in investment was directed to facilities on the University of Saskatchewan campus.

Saskatoon after 2005 led the province and nation in many economic indicators; perhaps most impressively, the city economy remained vibrant throughout the 2008–2011 global economic downturn (Table 1.2). Saskatchewan's provincial gross domestic product (GDP) per capita rose from 93% of the national average in 2001 to 105% of the national average in 2006 and, with a surge in commodity prices in 2008, Saskatchewan's per capita GDP surged to more than 30% above the national average. Given that

Table 1.2. Socio-economic indicators for Saskatoon and Canada.

Key indicators	Period	Saskatoon	Canada
Population change (%)[a]	2000–2010	14.0	11.2
Per cent foreign born[b]	2001	7.5	18.2
Per cent BA or higher[b]	2001	17.5	15.4
PhDs per 1000[b]	2001	10.9	5.4
Employment rate (%)[a]	2009	70.6	61.7
Unemployment rate (%)[a]	2009	4.7	8.3
Per cent creative occupations[b]	2001	33.4	29.2
Per cent S&T occupations[b]	2001	5.8	6.4
Bohemians per 1000 in the labour force[b]	2001	11.6	13.1
Number clusters[b]	2001	4	263
Per cent employment in clusters[b]	2001	15.5	22.1
Average household income (CAN$)[b]	2001	53,025	58,360

[a]Statistics Canada: http://www.statcan.gc.ca/
[b]Spencer and Vinodrai, 2006.
S&T, Science & Technology.

much of the new activity is centred in Saskatoon, this suggests that the local economy is delivering strong local returns. Moreover, Saskatoon has one of the highest labour force participation rates in the nation, with 71.7% of the potential labour force working, 4.7 points higher than the 2010 national average, and an employment growth rate of more than twice the national average since 2007.

While Saskatoon's economic prosperity has caught the attention of the national media and economic forecasters and pundits, a more fundamental and important transition has largely been overlooked. Saskatoon has converted from a town that overwhelmingly relied on brawn to a town that now increasingly defines itself as a city of brains. As the home for the largest provincial university and a major technical college, Saskatoon has historically produced a relatively large number of trained and skilled workers, but until recently the majority of graduates migrated to areas with more or better jobs (Calgary, Alberta, 600 km to the west, is jokingly referred to as Saskatchewan's third largest city, based on the migration of more than 120,000 skilled workers since 1945). Now, with better local economic prospects and stronger job growth, more graduates stay in Saskatoon and a few are returning. In 2006, Saskatoon had a higher stock of talent than the national average, with a larger proportion of workers with higher degrees (BA or higher) and PhDs (10.9 per 1000 of the population in Saskatoon versus 5.4 per 1000 of the population for Canada). Similarly, Saskatoon has a higher percentage of the population engaged in creative occupations (33% versus 29% for Canada). Somewhat counterintuitively, Saskatoon has a somewhat smaller share of its population in the science and technology occupations and a smaller share of its activity in clustered sectors. This new reality is slowly translating into a new image for the city. In the 1960s, the city described itself as the POW (potash, oil, wheat) city; in recent years, Saskatoon has both self-labelled and been singled out by others as 'Canada's science city', 'Paris on the prairies' and one of Canada's 'cultural capitals'.

There is debate about what underlies the economic dynamism in Saskatoon. Some argue that it is simply a recurring pattern of knowledge and capital accumulation driven by commodity markets. Others suggest that the infrastructure boom is a tangible, visible sign of a creative and innovative community. Saskatoon offers an excellent case to study the balance between these perspectives. As a small to medium-sized creative city with a strong innovation record (with world firsts in uranium and potash mining, farm machinery fabrication, crop varieties and biotechnology) flowing from a few world-class industrial clusters (especially mining and biotechnology), Saskatoon has been a major beneficiary of investments in soft infrastructure (research networks and projects) and hard infrastructure (e.g. buildings, laboratories and big science projects) funded by federal, provincial and municipal governments and agencies. Meanwhile, as a small Great Plains city, Saskatoon has until very recently had a predominantly white, midwestern look and feel about it, at least superficially suggesting that any success may be in spite of rather than because of economic, social and cultural diversity.

Saskatoon occupies a unique place in the list of cities in Canada. While Saskatoon's census metropolitan population in 2010 of 265,000 suggests that it has a number of natural comparator cities in Canada – Windsor, Victoria and Oshawa at over 300,000 population, and Regina, Sherbrooke and St John's at under 215,000 population – it is largely in a class of its own. Saskatoon occupies the top of the central place hierarchy in a market area of more than 600,000 (as a primary wholesale–retail centre), has no significant higher-order government and is relatively remote (minimum 600 km) from any city of competing or larger size (Stabler and Olfert, 1996).

STRUCTURE OF THE BOOK

The elephant metaphor we started this chapter with is fundamentally about getting a static picture of the beast. In practice, innovation is about change. A picture is a start, but it only captures a small part of the phenomenon.

This book offers an integrated, comprehensive assessment of the theories and their applications. The book explores the models, methods and metrics of innovation analysis in the context of a single centre – the global agri-food and oilseeds complex centred in Saskatoon, Canada. Over the past 10 years, a series of targeted yet complementary studies have been conducted on the Saskatoon economic region and its agri-food and biotechnology sectors. These studies, in combination, represent a unique examination of the application of innovation theory to a single region. The qualitative and empirical evidence from these studies provides a comprehensive picture of how innovation theory and evidence is supported, at least in the Saskatoon region, which may be a reasonable archetype for any advanced industrial agglomeration engaged in the global marketplace.

Chapter 2 reviews the main-line economic counterfactual for the more contextual innovation theories. The next six chapters review the theories, methods and evidence from applications in Saskatoon for the three competitive perspectives – Chapters 3 and 4 for place-based clusters, Chapters 5 and 6 for process-based innovation systems and Chapters 7 and 8 for people-based creative communities. Chapter 9 then investigates a range of new theories and methods that incorporate institutions, individuals and ideas in new ways. Chapter 10 takes what is known and assesses how it might better contribute to evidence-based policy making.

2 The Economic Explanation for Innovation

The most extensive literature about innovation, growth and development is based squarely in the neoclassical economic tradition and its extensions. The assumption in this literature is that rational, self-optimizing savers, investors, producers and consumers will allocate an optimal amount of resources to the innovation process that will generate a sustainable rate of technological change that enhances both individual and social welfare.

While much of the applied innovation literature and related policy advice deviates widely from the models and methods espoused by economic tradition, the base-line counterfactual to all contextual explanations of innovation remains the economic model. Economics has, for the most part, adopted the heuristic of Occam's razor as the rule of thumb that both guides the development of theoretical models and, in many ways, acts as an arbiter between competing models. Beinhocker (2006) asserted that this preference for parsimony can be traced back to the almost wholesale transfer of comparative statics from physics in the 1800s. There is strong evidence that economics adopted from physics the closed-system model, where change is an external disruptive shock. In that context, the most simple, straightforward or powerful explanation for the scale and scope of the shock is arguably the most appropriate one to focus on. As the rest of this book will show, innovation systems are never closed systems, and hence the simple comparative statics answers that the neoclassical framing offers is most often unproductively narrow and deterministic, triggering policy advice that is minimalist and not overly engaging for politicians and community leaders.

In order to appreciate the complexity of innovation and development, it is important to start at the roots of the theoretical model – the neoclassical paradigm. This chapter examines the classical framing of the economy (e.g. optimization, marginal analysis and the insights of the Ricardian trade model of comparative advantage, as modified by the Heckscher and Ohlin model/theorem; Heckscher 1919; Ohlin, 1933), reviews the resulting public choice policy prescription, investigates neoclassical growth theory, and introduces and discusses the implications of the new growth, trade and institutional theories. While all these perspectives purport to provide overarching insights into the formation and evolution of knowledge-based clusters, closer examination shows that the complementarities between the approaches are matched by clear differences in emphasis, which lead to some key points of disagreement.

THE MICROECONOMIC FOUNDATIONS OF INNOVATION

Generally, markets are assumed to be driven by self-interests. Individuals, operating as producers or consumers, sell or buy labour, other inputs, goods or services based on their own calculations of what is in their own best interest, mediated by market prices of the various factors and goods. These markets are driven by voluntary transactions. Unlike in the state or in social groups, there is no formal acknowledgement in markets that the interests of others are relevant or important. The economic theory of markets suggests that, under certain rigorous conditions, if everyone acts in a self-interested way, we can optimize our economic activities. Seabright (2004, p. 20) noted that the cooperation that delivers existing or new products often appears effortless because people and institutions embody a kind of tunnel vision, by 'play[ing] one's part in the great complex enterprise of creating the prosperity of a modern society without knowing or necessarily caring very much about the overall outcome'.

The concept of self-directing, anonymous markets has dominated much of the discussion since Adam Smith published his seminal book, *An Inquiry into the Nature and Causes of the Wealth of Nations* in 1776. This revolutionary document established the underpinnings of a liberal, individualistic-based economy, which ultimately helped to redraw the boundary between the state and the market. Smith did this quite deliberately, and his supporters, promoters and admirers similarly had an overt, political objective in promoting the concept of self-regulating markets as a way to govern both the economy and society (Sen, 1987). Prior to this, the feudal system dominated in many areas, with individuals having little or no choice in the production or consumption systems. While there had been some entrepreneurial development prior to this, the political system continued to constrain the role of the merchant class. By positing that self-interested individuals could lead to social and economic improvement, Smith provided another underpinning to the emerging liberal conception of government based on a social contract.

While Smith laid the foundations of an economy based on self-interest mediated by prices, others added the specific concepts that now comprise our understanding of the economic system. Over the succeeding 115 years, the theory was fleshed out and added to by a number of economic and social philosophers. Jeremy Bentham introduced the concept of utility and the notion of an economic calculus that could generate an optimal economy (based on a balance of individual valuations of pleasure and pain). William Stanley Jevons fleshed out the marginal utility theory of value. John Stuart Mill expanded on Bentham's concept of utilitarianism and added to our understanding of the economic underpinnings of bilateral exchange. Leon Walras elaborated on the concept of a general equilibrium and introduced the simplifying mechanism of the 'Walrasian auctioneer' to affect equilibrium prices and quantities. Ultimately, Alfred Marshall put it all together in a mathematical, graphical package with a deterministic supply and demand system that inexorably drives markets to an optimal output of goods produced and consumed, at one single set of market-clearing prices. While the validity of the underlying theories of supply and demand have been elaborated on and empirically tested over the past century (using an ever-increasing array of mathematical, empirical and computer tools), the basic theory of markets has been sustained.

Essentially, economic theory posits that individuals as producers, left to their own devices, will apportion their labour (either directly in own production or through selling their time to others for a wage) up to the point where the pain inflicted by the last unit of work equals the pleasure that could be derived from the output or wages one generates with that unit of labour. Similarly, individuals as consumers will buy increasing volumes of goods or services as long as the pleasure (i.e. utility) thus generated is greater than or equal to the cost of buying those goods. Individuals following this set of very basic, individualistic rules would thereby optimally allocate their own labour and income. Theoreticians have demonstrated that it is possible that an economy made up of such self-satisfying individuals could produce and consume an optimal amount of economic activity. This optimum is defined as the point at which maximum utility is reached – at that point, no better choice set can improve welfare, as working longer hours to consume more would be a loss in overall utility because the pain of the last unit of work would exceed the pleasure of the last unit of goods consumed, while working fewer hours would lower the pain but would lead to disproportionately larger declines in utility from consuming goods.

The production side of the theory was subsequently extended (by Jacob Viner and others) to convert the individual labour choice into a firm's cost structure. The theory suggests that, in most circumstances, the same qualities and quantities of one input (land, labour or capital), holding other inputs constant, will produce a smaller output for each additional unit of input. For example, adding more labour to a farm of a fixed size will eventually cause congestion and will relatively lower output per each additional worker. This concept of declining marginal productivity leads to each incremental unit of output costing more to produce. Graphed in a two-dimensional space, with volumes along the horizontal or x-axis and marginal value of product (usually converted to price) along the vertical or y-axis, the supply curve for any product by any producer is generally assumed to be upwardly sloping, as declining productivity of additional inputs leads to higher costs of producing a higher quantity of a good. The industry or product supply curve would simply be the sum of supplies offered by each of the producers at each price. The actual slope of the line is a function of the rate of declining marginal productivity – the faster the decline, the steeper the curve, and vice versa.

The consumption side of the economy was similarly fleshed out. The theory suggests that individuals will realize correspondingly less pleasure or utility from successively larger amounts of any good consumed. Declining marginal utility suggests that a consumer would probably be willing to pay a higher price for the first unit of a good they consume, but as successively more units of the good are consumed, pleasure or utility would decline and the price they would pay would correspondingly drop. In theory, at some point no further pleasure or utility would be generated by further consumption. Graphed in a two-dimensional space, with volumes along the horizontal or x-axis and marginal utility (usually converted to price) along the vertical or y-axis, the demand curve for any product is generally assumed to be downward sloping, as declining marginal utility would lead a consumer to be willing to pay less for each additional unit. The total demand curve would simply be the sum of all of the volumes demanded by individuals at each price level. The slope of the demand curve would depend on tastes, preferences, incomes and the prices of substitute and complement offerings.

Putting the supply and demand curves into the same two-dimensional space yields an X-like picture, with an upwardly sloping supply curve intersecting the downwardly sloping demand curve at some unique pairing of a price and a volume. The intersection point, or equilibrium, is optimal because no other points will lead to as much aggregate utility. If more output is produced than the optimum, consumers would only pay a price below what the producers' last unit cost to produce. Thus, while consumer utility could be higher, producers would be worse off. With voluntary exchange, lower prices would drive down production to the point where supply again equals demand. Alternatively, if a volume less than the optimum were produced, consumers would be willing to pay a price higher than the last unit cost to produce, leading to unusual profits for producers. Assuming there are no barriers to producing more, production would rise until the market clearing price equals the cost of the last unit. The process of finding a market-clearing price is not absolutely clear. Walras offered the idea of a conceptual auctioneer to manage what he called the tâtonnement process of groping towards the market clearing price – essentially, the auctioneer posts prices, seeking bids and offers, and only allows trade once a market-clearing volume and price combination is realized. If one accepts this construct, then there is little or no role for other governing systems – states and collective bodies could not improve on the choices of individuals. Seabright (2004, p. 21) argues that 'one of the great intellectual achievements of modern economics has been to work out very precisely the circumstances under which decentralized systems of market exchange can produce results that are efficient, in the sense of improving the condition of every individual as far as possible whenever this can be done without harming someone else ... the pattern of modern life has emerged without ever having been consciously willed by anyone'.

The end result of this theoretical construct is that there is no ongoing role for government intervention in the economy. Provided a few key conditions are met, anonymous individuals making self-interested choices will yield the optimal levels and distribution of economic activity. Economic theory and practice posit that governments are justified in intervening in the economy only if and when the market exhibits 'failure'. The difficulty comes in judging where and how to engage. The earliest and most morally consistent decision rule is the strict Pareto criterion, which asserts that the state should intervene only if it can make someone better off without making anyone else worse off (this is a type of economic Hippocratic oath to 'do no harm'). Given the nature of markets, however, there is virtually no intervention that will benefit some without disadvantaging others. Hence, moral philosophers have offered alternative criteria. Economist John Hicks and philosopher Nicholas Kaldor offered the notion of a compensated Pareto outcome, where in a democracy it would be acceptable if a state acts whenever it could compensate losers and still have winners; compensation would not need to be actually forthcoming. This rule is now used to define and justify a wide range of government interventions into the marketplace.

As one might anticipate, the underlying conditions for an optimal outcome are hard to realize – the so-called perfect-competition outcome is more of a theoretical construct than an observable reality. Theory clearly shows that it can only be realized if five rigorous and highly unlikely conditions hold. First, all actors must be atomistic – there must be a large number of small producers and consumers in a given market, each so small that their actions have no significant impact on others (i.e. both buyers and firms must be price takers and not exert any market power). Secondly, the goods and services must be homogeneous or perfect substitutes. Thirdly, there must be perfect and complete information – all firms and consumers must have complete information of production technologies, product attributes, prices and quantities. Fourthly, there must be no barriers to entry or exit for either producers or consumers. Fifthly, there must be no external effects of any production or consumption that lie outside the market – both positive externalities (e.g. spill-overs from research or public health) and negative externalities (e.g. pollution and congestion) must be known and included in the price of the product.

As different sets of assumptions are relaxed or break down, different types of imperfections arise, justifying different policy interventions. Many of these are important factors in framing innovation policy. Monopoly and oligopoly, where one seller or a small number of sellers respectively exert market power, occur when the first (atomistic) and fourth (entry) assumptions fail (sometimes caused by the absence of perfect information). We tend to see this in markets where there are physical, economic or legal barriers to entry (e.g. monopoly resource grants, large fixed costs to enter or patent grants). Monopsony and oligopsony, where there is only one or at most a small number of buyers who exert market power, similarly occurs when the first and fourth assumptions fail. Monopolistic competition, where there are many sellers producing highly differentiated goods, occurs when the second (homogeneous product) assumption is negated. Often governments will intervene with regulations, taxes or subsidies to limit the power of monopolistic entities or to encourage competition. To complicate matters, while market power is generally regarded as undesirable, as will be discussed below, it is often an important precondition for private investment in research and development.

When assumptions three (perfect information) and five (no externalities) fail, we face a set of market failures that are related to the inability to fully and efficiently transact to exchange goods and services. Coase (1960) argues that, with no transaction costs and full information, private bargaining should be able to handle any and all externalities. But transactions – when a good or service is transferred across a technologically separable interface, such as where one stage of processing or assembly activity terminates and another begins (Williamson, 1981) – are seldom costless. Transactions involve three cost components – search costs resulting from the efforts required to locate individuals with reciprocal interests; negotiation costs relating to efforts to conclude an agreement; and enforcement costs incurred in monitoring the performance of the parties to the transaction, which can be classified as either pre-contract (*ex ante*) or post-contract (*ex post*) costs. Furthermore, transactions may differ from one another depending on the extent of communication or strategic behaviour, the frequency of a transaction and the degree to which any asset involved in a transaction only has specific uses. The higher the transaction costs, the less likely we will see a perfectly competitive market outcome. Transaction costs are especially high for innovative products and services, as it is often hard to measure fully the costs and benefits of new things.

A variety of market-based institutions are often established to address market failures. Common ownership (e.g. vertical integration) is a common response to asset specificity; alternatively, some potentially opportunistic party could make *ex ante* credible commitments (often in the form of equity infusions or long-term contracts) in order to conclude transactions that are subject to significant asset specificity. Long-term contracting can also be used to address problems of opportunism. Sometimes, however, market failure cannot be addressed by private action. At times, governments can play the role of supra-firms, where their actions may reduce transaction costs; alternatively, collective institutions (sometimes referred to as non-governmental organizations or the participatory sector) may be more effective than government or the private sector. North (2005) asserted that a variety of innovations related to transactions were critical to the emergence of the modern economy. These innovations included increasing the mobility of capital (e.g. via the replacement of usury laws, the evolution of new financial instruments, such as the bill of exchange and development of accounting systems to monitor agents), lowering information costs (e.g. through evolution of printed prices, manuals on weights, measures, customs and brokerage fees, and integrated, efficient postal and telegraphic systems) and spreading risk (e.g. development of insurance, joint stock ventures and portfolio diversification). While many of these measures evolved through private initiative without any explicit government involvement, their transformative influence was only realized once the state entered into enforcement of commercial law. Regardless of the approach, however, each of these institutions (large, vertically managed firms, long-term contracts and various forms of government or non-governmental authorities) is usually governed using mechanisms other than the market. Supply and demand conditions do not operate effectively in these alternative configurations, with the result that prices cannot perform their allocative function.

Even if a competitive marketplace might equalize private benefits and costs, if there are externalities then the marginal social benefit might not equal the marginal social cost, creating a market failure that is often characterized as a public good, a common pool or a technical externality problem. Despite overall benefits from public goods (e.g. law enforcement or public health and safety laws), markets fail to produce adequate amounts as firms cannot charge users the full marginal cost because no one can be excluded from their consumption. Common pool resources (e.g. fisheries, clean air or market development investments) suffer from the lack of exclusion, and overuse can result in the depredation of the resource. Finally, technical externalities (e.g. pollution or knowledge) occur when activities of an individual can affect the welfare of others outside the transaction. The common policy response to externalities – especially related to basic research and development – is for governments to subsidize, invest in or directly undertake significant amounts of research.

This basic theoretical system has been internationalized over the last 150 years. Adam Smith acknowledged the logic and mutual benefit of international exchange wherever two trading countries had absolute advantages in the production of different products or services – if a country could produce a product more cheaply, it should specialize and export that product. Economist David Ricardo took that basic premise and through the now famous English wool–Portuguese port example demonstrated that, in a two-country world, even if one country has absolute advantage in all goods, there still would be gains for both countries to specialize and exchange. He posited that what matters more than absolute capacity is relative capacity – a theory that has come to be called the theory of 'comparative advantage'. In essence, the concept, formalized as the Heckscher–Ohlin theorem, states that a country has a comparative advantage and should specialize in the production and produce exportable surpluses of a product if the country is relatively endowed with factors of production (i.e. land, labour and capital) that are used relatively intensively in the production process. This theory clearly demonstrates that both amply and poorly endowed nations can gain from the efficiencies of specialization and trade. Further development of the theory showed that such specialization and trade will tend to benefit the locally abundant factors (the Stolper–Samuelson theorem), that the increase in one factor of production will cause relatively larger increases in the good that intensively uses that particular factor of production (the Rybczynski theorem), that trade in factors of production is a substitute for trade in goods (the Mundell equivalency theorem) and that trade over time will equalize the factor returns (the factor–price equalization theorem).

If one accepts the underlying economic logic, free trade between market economies delivers the best possible welfare for global society. If governments sought to support production of a good in their country that did not have any underlying comparative advantage (in that the local relative factor endowments did not match the relative factor intensities in the production of that good), then the investments would be for nothing in the long run as international competition and local demands for factors of production would eventually dissipate the subsidy (unless it was sustained in perpetuity). In essence, there is no strong rationale for government intervention in economic development.

This basic set of concepts has been a key driver for the liberalization of both domestic and international trade since the repeal of the Corn Laws in Britain in 1846. Surveys of economists suggest strong support for the notion that free trade 'is on the whole economically more beneficial than protection' – garnering support from 95% of US economists in one survey and 88% of economists in the USA, Austria, France, Germany and Switzerland. While Irwin (1996) showed that, in spite of this strong endorsement from economists, there has been strong and sustained pushback from many in industry, government and quite a few intellectuals, the preference for liberal trade remains the default advice from economists for governments around the world. Marshall (1923, p. 220) justified it best – free trade is an advantageous expedient because it 'is not a device, but the absence of any device. A device contrived to deal with any set of conditions must become obsolete when they change. The simplicity and naturalness of Free Trade – that is the absence of any device – may continue to outweigh the series of different small gains which could be obtained by any manipulation of tariffs, however scientific and astute.'

While the liberal market paradigm remains the orthodoxy, economists are willing to acknowledge that, in practice, the market involves a range of non-market, non-price mechanisms. This is most obvious when one broadens the focus to look beyond the allocative role of markets to its role in creation. Allocation is a static concept that addresses the optimization of the mobilization of inputs and distribution of outputs in a specified period of time. Creation, in contrast, is a dynamic concept that involves incentives for changing the current optimal distributions in order to expand future production and consumption possibilities. Systems that can handle static distribution often are unable to deal with dynamic challenges. While prices and non-price factors – including infrastructure underpinning the price mechanism, physical structures and organizational capacity of governments – are important for both roles, they pose a particular challenge for the creative activity. Generally, when markets are absent or underdeveloped, they cannot be expected to spontaneously create or generate any missing mechanisms.

GROWTH THEORY

For a long time, economists did not explicitly seek to extend the models to explain the accumulation of productive capacity and wealth – these were viewed to be largely artefacts of a properly functioning market, both in terms of efficient domestic supply and demand and international trade based on comparative advantage. This state of affairs continued until the 1950s, when the basic growth model was developed by Solow (1956). He posited that national growth is a function of the accumulation of labour and capital, with technological change an important but exogenous element to the model. The neoclassical explanation is that individual firms would respond to changing relative prices in factor inputs – i.e. labour, land and capital – and invest in innovating to minimize the use of the relatively scarce and expensive input (Hicks, 1932). This induced theory of innovation can explain quite a bit of iterative and sequential changes in technologies and industrial structures. While changes in relative prices are almost perpetual, the effects on innovative activity are assumed to diminish over time. Given that labour supply is largely a function of population growth, the only truly stochastic endogenous variable is capital accumulation, which is generally assumed to be a function of the marginal product of capital and the inter-temporal discount rate. The theory posits that the marginal product of physical capital declines as the ratio of capital to labour rises, so that the incentive to invest declines as an economy grows. At some point, the rate of capital investment should converge to a constant, with the result that long-term economic growth stabilizes at the rate of growth in the labour force. Both GDP levels and growth rates within and across nations should converge due to this process.

The evidence suggests that something is missing in this narrow specification of growth. Growth in per capita incomes has been sustained nationally and globally for long periods above the rate of growth in labour (studies suggest that the Solow model only explains between 20 and 50% of measured growth), and performance has varied greatly from country to country (Grossman and Helpman, 1991). Another deficiency of the Solow model is that it does not explain the role of firms in the growth process. Under perfect competition (a basic assumption of the model), firms are forced to price their products at their marginal cost, because their technology is completely transferable and new entrants will bid away any profits. Without the prospect of profit, there is no economic incentive to invest in innovating.

In response, an endogenous growth model was developed in the 1980s. Most of the new growth theorists start from Schumpeter's perspective that otherwise perfectly competitive firms pursue innovation to achieve monopoly profits during the time required for imitators to catch up. Schumpeter (1934) argued that, in practice, technological change is a strategic response by firms attempting to capture or create markets through product creation and differentiation. New products or new varieties of products create monopoly positions for the innovator, which allow the innovator to reap monopoly profits (called rents). But the existence of these rents creates incentives for other firms to imitate or innovate either to match or to leapfrog their competitors. Thus, monopoly rents from innovation are continuously under threat and likely to be of short-term duration. Schumpeter (1954) referred to this dynamic process as 'creative destruction'.

In this model, the focus is on the firm-based processes of investing time and other resources in the search for new technologies and processes. Grossman and Helpman (1991) argued that firms invest for two basic reasons – to reduce costs and to develop a new product that exhibits different quality characteristics (i.e. vertical innovation) or that provides variety (i.e. horizontal innovation). Regardless of the reason, innovators will continue to invest to innovate as long as they expect to earn a return on their efforts.

The new growth theory distinguishes innovations by two characteristics: rivalry and excludability. Rival innovations result in a good or service that can only be used by one person at one time (such as a consumer durable or personal service). Non-rival innovations involve an output (frequently knowledge, or what Romer calls recipes) that, for little relative expense or in some cases no cost, can be disseminated and used by every producer equally, whereby no one's use is limited by another's use. Excludability (sometimes referred to as separability) measures whether the innovation is protected from widespread use by legal means (e.g. a patent) or whether its adoption is limited by industrial structures or climate. If it is excludable, then the innovator can appropriate some or all of the benefits from the innovation. If it is not excludable, then the innovator is unlikely to be compensated for his innovation.

The traditional case of rival innovation, with or without excludability, typifies the Solow growth model, with decreasing returns to scale and ultimately slowing growth. As Grossman and Helpman (1991) observed, there is limited consumer demand, so that, as the number of product innovations rises, the average sales per variety will fall. Eventually, profit per innovation will stabilize and growth will converge to a stable path. They concluded that the stable rate of innovation ultimately is positively correlated with the taste for variety and the size of the economy, and that the efficiency of labour will be negatively correlated with the inter-temporal discount rate.

The more interesting case is where the innovation creates a non-rival product – either blueprints or applied science. If the firm that develops and owns the improved process acts like a pure monopolist and does not allow any other firm to use it (e.g. they do not license it), then that innovation would tend to exhibit decreasing returns to scale, as in the case of the rival innovation. Ultimately, it could stifle innovation and potential growth.

The key factor relevant to modern innovation theory is the non-appropriability of some of the benefits of innovation. Although economists have used a range of models to examine the effect of general or applied science innovations, the results converge on a common view. The new growth theory assumes that at least part of any non-rival knowledge is non-excludable. With technological change – described by Romer (1990, p. S72) as an 'improvement in the instructions for mixing together raw materials' – non-excludable knowledge spills over into the economy as a whole and raises the marginal value of new innovations. Theoreticians tend to assume that all innovation destroys the value of past innovations or investments. But there is also no reason to reject a priori the possibility that the externalities could improve the marginal productivity of existing capital and labour via more efficient production processes, especially if the innovation involves information technologies, which permit better management and new applications of existing technology. Although this would probably be a one-time upward adjustment in the marginal productivity, the adjustment would take time. Therefore, given continuous innovation, it is possible that growth could be bolstered over a long period. Hence, the positive externality associated with private investment leads to a sectoral or national production function with increasing returns to scale. In essence, the rate of growth in an economy rises with the amount of resources devoted to innovative activity (i.e. R&D, which is in turn a function of the size of the economy), the degree to which new technology is not excludable (i.e. the higher the degree of monopoly, the less innovation; conversely, the less it is excludable, the greater are the spill-overs) and a lower inter-temporal discount rate (i.e. the time horizon investors use for decision making).

Two aspects of this theory suggest that competing firms and industries will tend to concentrate in a few locations. First, if knowledge-based innovation involves extensive learning-by-doing, there would be extensive fixed costs of entering the industry. This tends to extend the period of monopoly profits. Assuming innovators are rational, they will recognize that, over time, their competitors will either innovate to imitate or leapfrog the current monopolist, thereby bidding down or eliminating the monopolist's source of market power and monopoly profits. So, innovators will be driven first to expand production and maximize profits during the period of monopoly and second to use some of these monopoly profits to continue to innovate to keep ahead of their competitors. Having monopoly profits allows the innovator to invest a greater amount in R&D and ultimately to widen the gap between it and the nearest competitor. Grossman and Helpman (1991) argued that, because each new innovation that increases variety or quality destroys the value of previous innovations, a monopolist would not innovate indefinitely. At some point (they suggest two steps ahead of their followers), the net present value of the investment in innovation would become negative. So it is possible that the next generation of a product might not come from the leader but from a close follower. If we assume generally competitive and efficient capital markets, markets would force this result. But with only limited market discipline over uses of retained earnings, it is possible – and often observed – that monopolists continue to innovate more than two steps ahead. The imperative to innovate has, in practice, tended to keep research and production units linked together in one or at most a few locations, in order to capitalize on the resulting synergies.

Secondly, although knowledge is a non-rival good among all producers worldwide, it might, at least in the short run, be excludable between jurisdictions for a variety of reasons. In the agri-food industry, for instance, climate, soil characteristics, microbiology and industrial structure all create natural or man-made barriers to transferring technology between jurisdictions. Some plant genetics and animals cannot survive or produce economically in different soil or climatic zones, and certain pests or microbes limit or curtail production for some crops and livestock, while many of the new genetically altered products require a certain scale of production unit (e.g. field size) or complementary investments (e.g. mechanized seeding, spraying and harvest equipment). Thus, it is possible and often observed that innovations in one country cannot be transferred elsewhere. The flip side of this is that like types of innovation will tend to concentrate in areas where there are similar climate, soil characteristics, microbiology and industrial structure. One result is that if the final product is tradable but the innovation-based knowledge is a non-transferable intermediate factor of production, then the fact that innovation begins in one jurisdiction could forever put that site on a higher trajectory of R&D and new product development. Grossman and Helpman (1991) argued that, as a result, the high-technology share of GDP and exports could be higher than otherwise.

The distribution of these gains from innovation is seldom left up to the chance operations of the marketplace. Although excludability is defined initially as the result of the attributes of the innovation, firms can improve the odds of gaining a larger share of non-excludable benefits depending on how they structure their operations. The evolving theory of institutional economics helps to define the potential for industrial structure to adapt to the market opportunities. Coase (1937) posited that firms exist to manage risk – namely those risks and uncertainties related to price discovery, negotiation and monitoring of transactions. Risk and uncertainty create costs. Risk in transactions can be managed if the market transaction is replaced by some institutional arrangement. Coase hypothesized that firms exist and operate because the cost of managing production in house is less than the cost of transacting to buy in. He concludes that firms will grow to the point where the cost of managing internal processes equals the cost of transacting with other agents.

This theoretical approach has been pursued by a number of researchers in recent years. There have been two key approaches: transaction costs and principal–agent theory. Williamson (1985) argued that contracting is not costless for two key reasons. First, markets are best described as operating with bounded rationality – that is, individuals act rationally but their options are limited by imperfect information. Secondly, he assumes that individuals and companies act opportunistically – that is, they act in a self-interested way 'with guile', by renegotiating terms of agreements to increase their returns. Their ability to succeed depends on their relative bargaining position, which is a function of the specificity of the assets each party has invested. The firm with assets that have little alternative use (e.g. science laboratories and grain elevators) are most at risk of having their returns bid away by other actors in the production system.

The alternative approach examines the costs and benefits of principal–agent relationships. The approach assumes that firms (principals) will contract with agents to avoid market risk. Once again, there is a concern that opportunistic agents will take advantage of any imbalance of power, in this case resulting from the inability to measure either their contribution to the total output (called non-separability) or their inputs (called programmability). The more measurement problems there are, the higher the cost of buying in relative to the cost of doing in house, with the result that vertical coordination is more likely to be pursued.

Mahoney (1992) put together the two institutional economic approaches to create a synthesized transaction cost–agency model. He argued that, if one assumes opportunism, one can predict the organizational form of vertical integration based on the degree of asset specificity, task programmability and non-separability. Many of these options are of interest for the agri-food industry. Crops-based agriculture traditionally has exhibited low task programmability, low non-separability and low asset specificity, so it tends to spot markets. But as the production technologies have become more linked (e.g. through proprietary, genetically modified, herbicide-tolerant seeds), task-programmability has risen (Phillips and Khachatourians, 2001). Recent efforts to breed specific market characteristics into seeds have increased non-separability. Given that asset specificity at the producer level remains low, these pressures should lead to more contracting in the industry. In contrast, in the genetics or seed business and in the related chemical industry, where asset specificity is very high, there is real pressure for vertical integration that enforces a more traditional hierarchical structure on the industry. The basic R&D sector, with low programmability and high non-separability, tends towards clans or communities (sometimes called innovation systems).

The model sketched above – with industrialized production chains innovating to develop market power – involves imperfectly competitive firms. The introduction of knowledge as a critical factor of production, which creates these imperfectly competitive firms, makes it possible that regional comparative advantage is now endogenous and not simply pre-determined by the relative endowments of labour and capital. In short, trade flows, and the resulting gains from trade, now have the potential to be driven by endogenous decisions and actions.

The impact of imperfect competition on trade volumes and composition depends on how far the technological externalities spread. If spill-overs are global, then relative endowments of traditional factors of production will ultimately determine trade flows. History might dictate the initial pattern of specialization as countries produce with the blueprints they inherit, but factor price equalization will drive trading countries towards the production pattern that fits its factor-based comparative advantage. So any impact of imperfect competition arising from endogenous growth would be only transitory. Grossman and Helpman (1991, p. 338) argued that technological spill-overs that are limited to a specific location (e.g. due to climate or industrial structure) create the possibility that 'comparative advantage is endogenously generated' because as 'countries engage in technological competition, comparative advantage evolves over time'. If technological spill-overs are geographically concentrated, initial and sequentially established conditions matter. In the extreme, if trade partners are similar in size and their endowments consist of a single primary factor, then a country that inherits even a small technological lead will come to dominate world markets for high-technology products. A productivity differential then becomes self-perpetuating. In more general circumstances, a large size, abundant human capital and a sizeable knowledge base contribute to a country's comparative advantage in research. In this case, there is an incentive for governments to subsidize research or to protect the local market for knowledge goods to provide an effective subsidy to a home producer (Brander and Spencer, 1985).

This case – representative of conditions in the large-frame aircraft sector and the large-memory computer chip markets – has been much studied by strategic trade economists. Baldwin and Krugman (1988), for example, showed that the actions of the USA and European Union (EU), and the US and Japanese governments, respectively, were rational, even though they have not been overly effective because of retaliation.

METHODS AND EVIDENCE

Some discussion of the applied research related to the theory is warranted. Testing the economic explanation for productivity, growth and technological change is largely the preserve of econometricians. Empirical work was initially constrained by computational limitations and incomplete data. While some growth data has been pieced together that spans about 140 years (mostly for the USA and western European countries), it is only since the 1960s that both constraints have been largely overcome, with a resulting profusion of large-scale analysis.

Most analysis starts with some technical specification of a simplified national production function that incorporates labour and capital. Cross-country, longitudinal econometric analysis has then been used to estimate the relationship and to identify the relative contribution of labour and capital to national and global growth. Specifically, economists have worked to test the theoretical hypotheses of declining marginal productivity and convergence.

Early estimates of the Solow–Swan model that examined growth in different countries and in different industries did reveal a reasonable fit but identified that a large portion of the estimated growth was not explained by the measured contributions of labour and capital. That so-called 'Solow residual' reflected both the impact of unexplained variables and total factor productivity. Mankiw *et al.* (1992) undertook to augment the model with accumulated human capital as a variable separate from labour, which improved the fit and brought the model more in line with the convergence theory. Barro (1991), another early analyst, undertook a study of around 100 countries, concluding that the models and evidence strongly supported the general notion of both the specification of growth and the hypothesis of conditional convergence. But even then, it was necessary to add other variables that are not fully justified in the reduced form theory of growth. His analysis showed that for a given starting level of real per capita GDP, the growth rate was positively correlated with higher initial schooling and life expectancy, lower fertility, lower government consumption, better maintenance of the rule of law, lower inflation and improvements in the terms of trade. Moreover, for given values of these variables, growth was negatively related to the initial level of real per capita GDP – in other words, wealthier countries are more likely to exhibit diminishing marginal productivity.

Another focus of analysis has been on the interrelationship between sectors and economies, especially through technology diffusion and spill-overs. Coe and Helpman (1995) reported in a study of 22 Organisation for Economic Co-operation and Development (OECD) countries that R&D spill-overs within and between those countries were substantial and Coe *et al.* (1997) showed that spill-overs from OECD countries to a large number of lesser-developed countries were substantial.

With the advent of the new growth theory of Lucas and Romer, there have been additions to the specifications of growth models to attempt to endogenize the investment in research and development, but the results remain quite general in nature. Overall, the policy advice flowing from classical and neoclassical growth theory and estimation is that effective markets, supported by appropriate incentives for investment in education and research and supplemented by some limited public investments in capital infrastructure, would be the optimal route for governments to follow – in effect, a light hand fixing market failures would be best (see Cooke, 2001, who explores the differences between the public and private investment models of the USA and the EU).

An alternative approach that has gained favour among some economists and opens the door to more contextualized analysis and policy was the move to comparative case studies and large-scale econometric analysis using panel data to inductively identify some of the unspecified but assumed important elements in growth. The ultimate goal is to identify underlying policy and institutional circumstances in individual countries and then to attempt to infer commonalities of their contributions to different growth rates. In this way, economists might be able to identify both necessary and sufficient conditions for growth. A recent example of this was the report by Acemoglu and Robinson (2008) for the Commission on Growth and Development, which examined 13 countries that recorded an average real growth rate of 7% a year or more for 25 years or longer in the period from 1950 to 2007. These cases revealed five 'striking points of resemblance': (i) they fully exploited the world economy through open trade, investment and knowledge systems; (ii) they maintained macroeconomic stability, with modest inflation and sustainable public finances; (iii) they mustered high rates of saving and investment and pursued a futures orientation; (iv) they let markets and market prices allocate resources; and (v) they had committed, credible and capable governments.

This recent move away from deductive, reductionist analysis towards more inductive, contextual analysis (albeit still mostly using econometric techniques) is a move towards the other literatures that focus their energies on the underlying structures, pathways and actors that engender and support innovation. Chapters 3–8 will examine these alternative approaches in more detail.

3 Clusters and Traded Interdependencies: Place-based Theories and Models

City regions are increasingly seen to be the engines of economic growth and innovativeness for the nation state. There is a strong debate about whether specialization or diversity is the key to success (e.g. Jacobs, 1984; Krugman, 1998a,b). One of the key questions in this debate is to what degree institutional factors (e.g. thick labour markets, industrial supply chain structures and physical infrastructure) generate success in the knowledge economy. Competing theories posit that success is based on the social dynamics and the economic, social and cultural diversity of a city region.

There is an evolving but incomplete set of theories about how local agglomerations, often called clusters, begin, evolve and operate, as well as a large and growing body of evidence about a range of actual or potential clusters identified by policy makers and analysts. The term cluster is a bit of a flexible friend. It is used widely both in the academic literature and among economic development practitioners. The *Oxford English Dictionary* defines the generic term as a 'group of similar things, esp. such as grow together'.

While the phenomenon of firms and various not-for-profit entities in the same sector or product market tending to co-locate in specific geographic regions (rather than spreading out evenly across the geography or economy) has been observed since the beginning of recorded economic history (e.g. Braudel, 1973; Boorstein, 1983), the active study and search for ways to promote that outcome has only been engaged more recently. Economists first began to develop models to explain agglomerations of like entities in the 1700s (e.g. Smith, 1776), and by the mid-1800s economists were beginning to develop new explanatory theories and undertake intensive analyses of the phenomenon. Given the antecedents of this approach, the general view

Innovation in Agri-food Clusters: Theory and Case Studies (P.W.B. Phillips *et al.*)

was that localized agglomerations were more of an artefact of the general functioning of competitive, openly trading economies, based on subtle factors of cost economics. Most economists shied away from translating this into proactive industrial policy advice, being concerned that politicians and bureaucrats are unlikely to be able to make better investment choices than individual entrepreneurs and firms.

Nevertheless, a counter-movement led by scholars drawn from a number of newly industrializing countries developed the 'infant industry' argument in support of protective tariffs and proactive industrial strategies. The core of the argument is that new firms and industries often do not have the capacity to compete with established competitors in more advanced industrial economies and thus need to be protected from competition until they can attain similar economies of scale (this logic resonates with the work of Brander and Spencer, 1985, and Krugman, 1986, except that, in the more recent case, government intervention was only warranted if the industry exhibits increasing returns to scale over a wide range of volumes). The infant industry argument was initiated in the USA in 1790 by Alexander Hamilton, in his *Report on Manufactures*, and then was translated and transferred to Germany by Friedrich List following his visit to the USA in the 1820s. List formalized his arguments in 1841 in *The National System of Political Economy*. Since then, many countries have sought to industrialize behind tariff walls or with the support of special subsidies. While some countries argue they were successful in generating local economic activity that would not otherwise have been there, it is not clear that the true infant industry exhibits the attributes of a cluster – rather it is more a function of the specific influence of the policy and at times administrative direction on firm location and structure.

While that debate continued in fits and starts into the 1900s, the rise to dominance of the neoclassical economic paradigm after 1950 pushed studies of proactive industrial development (and related policy prescriptions) to the margins of the economics literature. The neoclassical social consensus among academic economists and practitioners (in governments in advanced industrial democracies and at the World Bank, International Monetary Fund and the General Agreement on Tariffs and Trade (GATT)/World Trade Organization) was that governments should only intervene in markets in the case of extreme market failure. Intervention was contingent upon the identification of a market failure and demonstration that intervention had a strong probability of remedying the imperfections in the market. In essence, the neoclassical social consensus was that markets should be left to their own devices.

The winds began to change in the 1990s. Beginning in the early part of the decade, economists began to refocus their attention on the microeconomic foundations of growth. After a decade of stagflation, new conservative governments shifted to a low-inflation, tight-money macroeconomic stance and began to look for new microeconomic options to accelerate productivity and economic growth.

Michael Porter's well-timed release of *The Comparative Advantage of Nations* in 1990 reintroduced the concept of clusters, this time couched in a paradigm that posited that local competition is the primary dynamic behind cluster development and sustainability. This concept worked well with the shift in strategies on the part of governments. Since then, the general concept of similar things growing together has been applied widely to economic and industrial policies around the world.

THE THEORETICAL BASE FOR CLUSTERS

Cluster theory is a blend of many threads drawn from economic geography and regional economic theory. While there is no consensus on the full and complete theoretical explanation for clusters, there are a few threads that are common to most explanations of the phenomena. Expressly, spatially specific, traded interdependencies are at the root of the cluster stream.

Much of that work returns to the basic observations by Marshall (1890), who identified three clear and straightforward sources of external economies (Krugman, 1998a, calls them centripetal forces) that explained the co-location of some types of industry. His first of three explanations that external economies develop from informational spill-overs is the most complex and difficult to measure:

> When an industry has thus chosen a locality for itself, it is likely to stay there long: so great are the advantages which people following the same skilled trade get from near neighbourhood to one another. The mysteries of the trade become no mysteries; but are as it were in the air, and children learn many of them unconsciously. Good work is rightly appreciated, inventions and improvements in machinery, in processes and the general organisation of the business have their merits promptly discussed: if one man starts a new idea, it is taken up by others and combined with suggestions of their own; and thus it becomes the source of further new ideas.

Secondly, he noted that some centres developed an extensive local and regional network of related and supporting industry (often called 'backward linkages'), as well as access to large, sophisticated markets ('forward linkages'), which makes that location more attractive for investors:

> And presently subsidiary trades grow in the neighbourhood, supplying it with implements and materials, organising its traffic, and in many ways conducing to the economy of its material. Again, the economic use of expensive machinery can sometimes be attained in a very high degree in a district in which there is a large aggregate production of the same kind, even though no individual capital employed in the trade be very large. For subsidiary industries devoting themselves each to one small branch of the process of production, and working it for a great many of their neighbours, are able to keep in constant use machinery of the most

> highly specialised character, and to make it pay its expenses, though its original cost may have been high, and its rate of depreciation very rapid.

Finally, he argued that, as the size of the labour market grows, it begins to specialize and support further development:

> Again, in all but the earliest stages of economic development a localised industry gains a great advantage from the fact that it offers a constant market for skill. Employers are apt to resort to any place where they are likely to find a good choice of workers with the special skill which they require; while men seeking employment naturally go to places where there are many employers who need such skill as theirs and where therefore it is likely to find a good market.

Clearly, these are not the only forces working on industrial location. If they were not offset by centrifugal forces, all industries would tend to specialize and produce in only one location and there would be extensive trade between locations. Krugman (1998b) proposes three offsetting forces (Table 3.1). First, agglomeration tends to create congestion, which is a pure diseconomy of scale. Secondly, immobile factors – land, natural resources and, internationally, labour – at times force production to go where the resources are. Finally, as economic activity in a region expands, it increases

Table 3.1. The forces of agglomeration. (From Krugman, 1998b.)

Centripetal forces	Centrifugal forces
Labour force externalities • Specialized skills • Low search cost to fill positions • Inter-firm flow of employees	*Immobile factors of production (not available)* • Natural resource inputs • Lack of labour and/or skills • Barriers (e.g. distance) to market access
Non-traded input externalities • Availability of financial capital • Business support services (e.g. consultants or advisors) • Cost of essential services (e.g. fire, sewerage, transport, waste removal)	*Land rents and other costs of doing business* • Rents • Taxes • Cost of services
Knowledge flow externalities • High-quality personnel (e.g. development and availability) • Stock of ideas and idea production environment • Track record of interaction and collaboration • Informational spill-overs between firms	*Pure external diseconomies* • Pollution • Crime • Traffic

demand for land, which is relatively fixed in supply, tending to drive rents higher.

While clusters often are characterized as having multiple categories of actors, profit-motivated firms are unambiguously the key. Their search for Schumpeterian-style market advantage creates an inexorable drive to invest and search for new processes, products or organizations. As they pursue new opportunities, they operate as agile, fast-growth enterprises; if the innovative effort matures, they will almost inexorably go through a variety of stages ranging from competitive firms to oligopolies and ultimately to mature or declining firms (Table 3.2). The more innovative the firms, the more attracted they are to skilled labour pools or potential collaborators or competitors. As firms mature, their attachment to specific research locations can weaken to the extent that relocation may be feasible. For that reason, regions desiring to build sustainable industrial capacity are increasingly driven to support and nurture sustained innovation as a fundamental part of any policy.

Starting from Porter, one can see at least three different literatures that fit within the typology of the cluster: (i) managerial focused cluster analysis; (ii) clusters embedded in supply or value chains; and (iii) the role of anchoring institutions in clusters.

Porter (1990) clearly focuses on the role of firm management and strategy to drive development and competition in localized industry. His graphic representations of concentrations of competing and cooperating companies, suppliers, service providers and associated institutions amply illustrate the industrial interdependence and managerial practices through

Table 3.2. A neo-Schumpeterian model of industrial development. (Adapted from Lundvall, 1992.)

Factor	Innovative stage	Competitive stage	Oligopolistic stage	Declining stage
Localization pattern	Close to skilled labour pools or founder's residence	Firms attracted to least cost site	Relocations retarded for early firms but encouraged at later stages	Relocation from old plants to modern plants in new regions
Importance of proximity	Agglomeration economies high	Proximity declining	Firms operate in larger input and output markets	Low
Growth	High	High	Low	Negative
Technological development	Innovation is key	Products are standardized; economies of scale	Product differentiation and process development	Limited; profit maximization dominates

supply-and-demand relationships. While Powell (1990) characterizes these relationships as networks, he follows in Porter's footsteps, focusing on the flexibility of markets and the visible hand of organizational authority. Maillat (1988, 1991) talks about these locations as innovative milieu, defining them as 'an organization, a complex system made up of economic and technological interdependencies ... a coherent whole in which a territorial production system, a technical culture, and protagonists are linked'. Rosenfeld (1995) goes farther into examining the complex system, defining agglomerations as highly geographically concentrated groups of companies that either work directly or indirectly for the same end market, share values and knowledge so important that they define a cultural environment, and are specifically linked to one another in a complex mix of competition and cooperation.

The second stream involves a wide range of scholars and practitioners who have looked at co-location and the resulting specialized forward and backward linkages as conferring performance advantages (e.g. Doeringer and Terkla, 1995). This stream quite explicitly defines agglomerations in the context of industry clusters or well-defined supply or value chains. Gollub (1997), for instance, asserts that these closely linked, interdependent firms in a supply chain drive the economy by creating 'commuter sheds' that supply inputs sufficient for companies to maintain competitiveness, generating seven factors that feed intra-cluster activity (i.e. technical resources, human resources, financing, physical infrastructure, advanced communications, a receptive business climate and quality of life) and deliver secondary jobs (what he calls 'Mom and Pop' jobs) that provide a penumbra of supplementary employment and income opportunities that buttresses the cluster. Others are more focused on transactions, examining the explicit buyer–seller relationships that are documented and analysed through quantitative input–output matrices (e.g. Enright, 1998, 2000; Roelandt and den Hertog, 1998; Bergman and Feser, 1999). Some look more at a specific product or market, which is then often characterized as a network of strongly interdependent firms, which are correspondingly linked both to knowledge-producing agents and to customers in a value-adding production chain (Roelandt and den Hertog, 1998). Bergman and Feser (2004), for example, explicitly examined the cluster as an extended input–output or buyer–supplier chain, sometimes involving multiple sectors or industries, but always including final market producers and first-, second- and third-tier suppliers that directly and indirectly engage in trade. Maskell and Lorenzen (2003), in contrast, focused on the 'swollen middle', analysing the comparative institutional structures and the varied roles of markets as institutions.

A third approach concentrates instead on specific actors in the cluster. The European Association of Development Agencies (2004) explicitly identified and focused on hard and soft infrastructure (e.g. universities, venture capitalists and other spatial assets), assessing their effect on the critical mass of information, qualifications and relationships in a given sector. They posited that each country or region enjoys specific local conditions and common customers and partners that link companies through trading relationships, thereby nurturing and sustaining commercial competitiveness. Rugman and D'Cruz (2000) more explicitly investigated the role of flagship firms, usually multinationals, which bring scale economies and extensive forward and backward supply chain relationships and thereby anchor clusters by providing global reach into knowledge systems and foreign markets.

While the focus of each of the streams varies from the macro (management) to the meso (the value chain) to the micro (specific infrastructure or anchor firms), all of the scholars agree that any resulting interdependencies are for the most part traded. It is important to note that the existence of a cluster can at best be treated as a hypothesis in terms of examining concentrations of economic activity in the global agri-food sector – its existence and function should not be assumed but rather should be tested as a possible hypothesis.

COMPETITIVENESS AND THE PLACE-BASED PERSPECTIVE

The narrow cluster theory above lacks a clear causal pathway for the three identified locational factors to influence firms. Given that place-based cluster theory posits that these factors are absorbed into firms through a combination of active strategic management and transactions that access the interdependencies, it is somewhat surprising that most of the theories treat the firm like a black box – firms are simply assumed to exhibit these motivations and to undertake these functions without any clearly defined explanation of why or how.

In response to this weakness, a number of business theorists, in what has come to be called the 'competitiveness perspective', have articulated a model whereby proximity fosters direct economic linkages characterized by shortened communication lines, more and better intelligence related to competitors and markets, lower transaction costs (e.g. search, negotiation and enforcement) and intensified rivalry. All of these directly or indirectly relate to explicit or implicit firm strategy. Direct economic linkages are assumed to be embedded in stable exchange relationships among and between firms and their suppliers, distributors, competitors and customers (Gordon and McCann, 2000). These are often referred to as backward, forward or horizontal trade linkages.

This perspective is most often associated with Porter's (1985, 1990, 1998) work on competitive strategy. This work is not explicitly about clusters, but it did identify industrial clustering as a driver of industrial and national competitiveness and contributed to moving economic geography from the periphery to the mainstream. The essence of Porter's work is that national competitiveness is an extension of industrial competitiveness; industrial competitiveness is, in turn, determined at the regional level, the location and action of a range of agents. Agents could include government organizations, industry associations, business service providers (e.g. accountants, financiers and management consultants) and other service organizations. Porter introduced a 'diamond' model to categorize and assess the competitive advantage of nations, particularly in the context of firms and their relationships. The diamond assumes clusters entail: (i) a primary role for firms, categorized by their strategy, structure and rivalry; (ii) knowledgeable and demanding consumers; (iii) a range of relationships between firms and an array of related supporting industries; and (iv) a range of general use and specialized factors of production, such as universities and infrastructure.

From industry to industry and region to region, collections of cluster agents will look quite different. But what does not change is the driving force behind how the agents are related. According to the diamond model, each agent has some form of exchange relationship with the core competitors in the localized industry. As economic exchanges, they will only be established and sustained when some value can be derived by both parties.

Exchange relationships ought to be viewed as one-time interactions between two parties, where each interaction is assessed on a strict cost–benefit basis. According to the location theory, these exchange relationships may lead to clustering as several firms within an industry 'attempt to minimize distance, transportation, and production costs; obtain cheap labour; and minimize risks' (Dicken and Lloyd, 1990, as cited in Bekele and Jackson, 2006, p. 4). From the competitiveness perspective, exchange relationships are important for more than just the case-by-case benefits they generate. Exchange relationships form the basis for sustained linkages or interactions to emerge. Linkages can be viewed as conduits for recurring exchange. As such, linkages become an invisible infrastructure through which a variety of benefits can flow. Porter indicates that the initial exchange relationships are important to the emergence of the cluster, but it is the maintenance of trade linkages that fosters long-term competitiveness.

Short-term benefits from local exchange relationships can promote co-location. Benefits can include lower search costs and lower transportation costs. Clusters persist as firms work to maintain exchange relationships and foster an invisible infrastructure that enables additional benefits such as access to information about competitors, awareness of new innovations and a greater sense of rivalry. If these exchange relationships are entrenched as sustained linkages, then second-order benefits (e.g. market and competitor intelligence, technology dispersion and intensified rivalry) become important. These benefits extend beyond the short term and, as Porter emphasizes, are critical to accelerating the innovation process and enhancing competitiveness. Porter cautions, however, that co-location can also foster negative effects such as ebbing rivalry or regulatory inflexibility that can ultimately lead to the exhaustion of agglomeration effects.

Therefore, the central argument underpinning clusters is that proximity converts short-term benefits from exchange relationships into second-order benefits that impact long-term competitiveness. The question then is how firms position themselves to realize these benefits.

CLUSTERS IN PRACTICE

No matter how vague the term and the underlying causal pathways, policy advisors and practitioners have widely adopted versions of Porter's cluster. Economic development agencies in developed and developing countries have retooled the infant industry argument through the cluster lens and applied the generalized approach to their communities, customizing it to their particular geopolitical region. Location is increasingly becoming the focal point for discussion of the dynamics and benefits of regional development, especially for knowledge-based development related to the application of new sciences, such as agricultural biotechnologies in the agri-food sector.

A survey of the recent application of cluster formation and operation (Table 3.3) illustrates the scale and scope of its use. While Porter (1990) has done the most comprehensive and complete survey of clusters, one can see a range of approaches to understanding and promoting clusters as a development model. Taken together, the studies cited Table 3.3 encompass more than 3200 examinations of clusters or nascent clusters, undertaken by scholars of economics, business and economic geography, by practitioners in consultancies and by policy advisors in regional, national and global development agencies. The studies range from conceptual investigations to individual and small sets of case studies and ultimately to full-blown standardized international comparative analyses. The approaches varied significantly. The individual and small sample case studies used a mix of highly contextualized narrative methods while others used well-defined quantitative and qualitative frameworks. The sample studies almost all used clearly defined frameworks that offered the maximum opportunity to codify quantitatively or qualitatively the structures and outputs of clusters in order to facilitate cross-cluster comparisons. Finally, the focus of cluster studies varies widely. Academics and global development agencies, such as the OECD and the World Bank, tended to draw their sample from regions in a variety of countries from multiple continents. In contrast, others focused on investigating their membership (either internationally or nationally), which usually limits their analysis to clusters inside the group. Similarly, some scholarly groups and many governments and consultancies focus exclusively on the lessons one can learn from their constituent clusters.

Table 3.3. Alternative approaches to cluster analysis (chronological). (Adapted from Ryan and Phillips, 2004; Phillips *et al.*, 2004.)

Author/institute	Title	No. of clusters	Analytical approach	Cluster type(s)	Scope
Porter	*The Competitive Advantage of Nations* (1990)	2500+	Framework	Industrial	International
KPMG, DRI/McGraw-Hill and IMPAX Policy Services International	*Building Technology Bridges: Cluster-based Economic Development for Western Canada* (1996)	Three interprovincial clusters	Explorative/ framework	Agri-Food, forestry, and energy and mining	Western Canadian region (BC, AB, SK, MB)
National Research Council Canada	*Building Technology Clusters Across Canada* (1996)		Institutional based	High technology	Canada
Atlantic Canada Opportunities Agency (ACOA)	*Prospects for Growing Knowledge-based Industrial Clusters in Atlantic Canada* (1997)	Six (three pan-regional/three regional)	Framework	IT, geomatics, oceans, medical devices, food processing	Atlantic Canada
World Bank LED (Local Economic Development)	*Cluster-based Economic Development: A Key to Regional Competitiveness* (EDA, 1997)	17+ case studies	Framework	Non-specific	Global (developed)
OECD Focus Group on Industrial Clusters	*Cluster Analysis and Cluster Based Policy in OECD Countries: Various Approaches, Early Results and Policy Implications* (1998)			Multiple sectors	Global (developed)
Scottish Enterprise Network	*The Clusters Approach* (2000)	3 (national/ industrial focus)	Prescriptive/role of HEIs/framework	Biotechnology/food and drink/semi-conductors	Scotland
Ryan	*Overview of International Life Science Clusters in Australia, the EU and North America* (2001, unpublished report)	20+	Framework	Life science	International (Anglo Teutonic)

Table 3.3. Continued

Author/institute	Title	No. of clusters	Analytical approach	Cluster type(s)	Scope
Institute for Strategy and Competitiveness, Harvard Business School	*Findings from the Cluster Meta-Study* (2002)	664 (developed nations) 169 (developing nations)	Application of Porter framework	Non-specific	Global (developed and developing)
Global Cluster Initiative Survey	*The Cluster Initiative Greenbook* (2003)	500+	Framework and practitioners' guide	Non-specific	Global (developed)
The Competitiveness Institute	Clusters Initiative Database (as of 2003)	81	Framework	Industrial	International (membership based)
Clusters Asia Pacific Inc.	Clusters Asia Pacific Inc. (as of 2004)	35 paid membership with a network of 700	Virtual practitioners' network	Industrial	Asia Pacific (membership/ network based)
UK Department of Trade and Industry	*A Practical Guide to Cluster Development* (2004)	Unspecified	Framework (multinational)	Multiple sectors	Applicable for UK
ISRN	Clusters Initiative (2000–2011)	26	Narrative	Industrial/ Innovation/ High Technology	National (Canada)
Scottish Economic Policy Network	*Scottish High-technology Clusters: an Empirical Analysis of Firm Linkages, Innovation and Performance* (on-going)	Unspecified	Inter-firm analysis	High technology	Scotland

Looked at in a different way, one can see differentiation in terms of the models used to analyse clusters in various countries throughout the world. For instance, the US work tends to focus on commercial outcomes and investment attraction and to emphasize the role of the academic institution (in terms of innovative and patent outputs) in the regional cluster. In Europe, there is a marked emphasis on the social factors associated with clusters. The public sector is deemed to drive the process of clustering. Canadian analyses tend to be concerned with exploring innovation in terms of cluster-based models and approaches, while Australia appears to emphasize the role of seed funding and anchor organizations at the cluster level.

There are a fair number of agri-food-related clusters embedded in these studies, but it is quite difficult to draw any lessons from them in their integrated form. There have been two cases where the role of clusters in the agri-food setting has been explored.

Phillips *et al.* (2010) delved into studies examining the key agri-food clusters that have been studied by others (Table 3.4). They then took eight of the cases to study in more detail and offered some insights and lessons into these types of clusters:

- BioBelt crosses multiple states and appears to have both extensive MNEs and small and medium-sized enterprises and start-up activity that is generating new products that add value to farmers and the agri-food sector.
- Western Australia has used national R&D strategies to strategically position itself at the centre of a global network of R&D related to pulses and other small grains and oilseeds, thereby leveraging international capacity.
- Food Valley is a leading EU producer of new crop varieties, built on a base of public research and public–private partnerships.
- Thise in Denmark has taken heavily regulated milk markets and found a way to develop higher-value innovative product lines.
- New Zealand has repositioned itself through extensive state-industry efforts as a global leader in kiwi fruit and terroir wines.
- Central Scotland has integrated its industry and research actors into producing safe, high-quality, premium-priced Scottish beef.
- Queensland has had a decade old strategy of improving meat quality and expanding international markets by moving up the value chain through genetics.

Each offers the opportunity to identify different institutional perspectives and both strategic and tactical policy lessons. The eight cases offer interesting vignettes of different mixes of policy tools. While good economic and price fundamentals are arguably in the background of the success in the eight cases, each highlights a different mix of policy actions. In some cases, governments focus on nurturing centripetal forces in specific places (e.g. the US BioBelt and the New Zealand wine cases), many focus on nurturing specific institutional supports for innovation (e.g. of public–private partnerships in the Australian

Table 3.4. Agri-food cluster case studies. (From Phillips *et al.*, 2008, 2010.)

Study	Agri-food clusters
Porter (1990)	Denmark: agricultural machinery, dairy products, food additives Italy: wool fabrics Sweden: semi-hard wood flooring Switzerland: chocolate, confectionery UK: biscuits USA: Agricultural chemicals
ISRN studies (2000–2010)	Saskatoon oilseeds cluster/innovation system Okanagan wine cluster Niagara wine cluster
The Competitiveness Institute (Cluster Initiative Database)	Croatia: Croatia wood cluster Dominican Republic: banana cluster, Jarabacoa coffee cluster, Jarabacoa greenhouse cluster, PROMANGO (mango) France: Cereales Vallee (grains), Cluster des 'bio produit' (organic cluster), Qualitropic (overseas agri-nutrition in the tropics), Toulouse Agri Campus (agriculture and veterinary), Vegepolys (horticulture) Georgia: tea cluster development Mauritius: Chemin du Sud (grassroots entrepreneurial) Philippines: industry clusters in the Caraga Region (wood, oil palm, abaca, vegetables, seaweed) Serbia: South Serbian Alko Cluster (grape, fruit, wine, brandy) Turkey: Bartin Agriculture Cluster (urban agriculture), Urfa Organic Agriculture Cluster (organic farming) USA: Maumee Valley Growers (greenhouse), Monterey Agriculture Cluster (crop production, livestock, food processing, agricultural machinery and equipment, agriculture related transportation and distribution)
Ryan and Phillips (2004)	Canada: Innovation Place – Saskatoon (canola, flax seed), Agri-food quality cluster – Guelph (corn, miscellaneous) USA: Connecticut Bioscience Cluster (corn, fruit), The Research Triangle Park – Raleigh, North Carolina (corn, soybean), BioBelt – St Louis (corn, soybean), San Diego (forestry, fruit, vegetables) Europe: Innovation Triangle – Scotland (livestock, animal cloning, potatoes), Bio Valley – France, Germany, Switzerland (cereals, cotton, corn, livestock) Australia: Qbio – Brisbane (forestry, aquaculture, horticulture), BioHub – Sydney (livestock, cotton, pulse, wheat, canola, food processing)

pulse industry and the Scottish beef case and the core partnerships in the Danish organic industry and the French cheese system) and a few target enhancement of the role of individuals (e.g. entrepreneurs in the Dutch Food Valley and the co-regulators in the Australian beef food safety and quality system).

In an earlier, complementary study, Ryan and Phillips (2002) investigated a cross-section of biotechnology-focused clusters. They concluded that most biotechnology-focused clusters, regardless of location, tend to centre R&D activities on health care, diagnostics, the environment or pharmaceuticals. Agriculture-related biotechnology activity, particularly in the European and Australian clusters, seems to operate as a subset of the biotechnology subsector as a whole, leveraging complementarities with the health- and environment-related supply chains and infrastructure within the given region. The only readily identifiable, dedicated agricultural biotechnology clusters with scale are in North America – the Saskatoon-based cluster is one of the most advanced agriculture-dependent examples.

Ryan and Phillips concluded that definitions of clusters varied widely from region to region and country to country, depending on the actors and institutions involved and the strategies that they employ. The national regulatory environments and intellectual property rights regimes add to the complexities of a given cluster, further complicating the characterization process. While the focus of cluster policy is for the most part on exploiting traded interdependencies, most studies identified a range of factors, such as trust, habits and conventions, that play a supportive role in localized learning and in the flow of codified and tacit knowledge. The highly contextual nature of clusters makes them especially difficult to measure and analyse – they are, in many ways, characterized more easily by their differences than by their similarities. Based on available research, Ryan and Phillips concluded that practitioners characterize agri-food and biotechnology clusters less based on measurable scale and more on: (i) their ability to self-define; (ii) the markets they focus on; (iii) the number of world firsts or product firsts that they introduce to the global market; (iv) the existence of comprehensive government and promotional strategies; (v) the existence of demand-driven knowledge transfer structures; and (vi) the influences of geopolitical structures and the environment.

METHODS

As one might expect given the wide range of interpretations of clusters, scholars and practitioners have used a wide range of methods to describe, delimit and test the boundaries and impacts of industrial agglomeration.

At root, there are a number of important considerations behind the study of place-based clusters. In the first instance, the bulk of the work is fundamentally interdisciplinary analysis, which generally is not derived from theory alone but rather is most often triggered by observations of society or nature. Thus, the very nature of defining clusters as the problem area tends to be a normative exercise. Moreover, the structure and focus of the analysis tends to be grounded in the definition of the problem, rather than in any particularly coherent or specific methodology. This leads to a problem of finding a clear scope for the analysis and a single, generally acceptable methodology for undertaking cluster work.

While many critics assert that this leads to a lack of method, this is probably an unfair complaint. The *Oxford English Dictionary* defines method simply as a 'special form of procedure, esp. in any branch of mental activity'; methodology, thus, is 'orderly arrangements of ideas; orderliness, regular habits'. Method or methodology will thus vary widely among different disciplines and approaches, as they each use a different array of postulates, rules and data. Because the method chosen is often inextricably entwined with a discipline and a researcher's approach, it is often a hidden or overlooked aspect of many areas of study. Given that clusters have been studied by such a wide range of disciplines, one can see that the choices the researchers and analysts make about a research topic – why they want to examine the topic, what aspects they will examine, how they will examine it, and what data and evidence they will use to advance their arguments – often predetermines the results and affects their ability to accurately reflect probable events or to influence debate.

This brings us back to the blind wise men. Ultimately, what researchers and policy analysts examine and what they conclude is heavily influenced by the underlying motivations of the researcher. Some investigators and disciplines are inherently normative, in that they are driven by an ethical or normative judgement of 'what ought to be'. They often start with a perspective of wanting to change the world, and then identify the area of analysis and define the methods in ways that further that goal. As such, normative studies are frequently challenged as being biased and incomplete. Much of the work done on clusters seems at least partly to reflect a normative bias. Many of those who have investigated the topic appear to have a driving concern to affect the distribution of economic activity in ways that would not be realized under the neoclassical paradigm discussed in Chapter 2. While this is not a fatal flaw, one needs to be aware that it can divert attention from a full, clear and complete examination of the underlying factors of economic activity and instead focus attention and effort more on identifying and examining options for change.

In contrast, many disciplines and investigators strive to remove the ethical and normative perspective from their work, instead seeking to pursue an objective, positivist examination of 'what is'. Economists have unambiguously adopted this approach. This reductionist approach allows investigators to test specific aspects of clusters, such as the role of anchor firms, while holding the whole constant.

Another area of disconnect is in the choice of research approach. Many people distinguish between two basic kinds of logical argument: induction and deduction. Induction is usually described as moving from the specific to the general or, more particularly, it is the process of arriving at a conclusion based on a set of observations, which involves finding patterns from a complex environment. Inductive arguments often include analyses that assume that the whole is like the individual parts, facilitating extrapolations to areas beyond the study (which are assumed to be like those studied) and forecasts based on the past. Deduction begins with general theory, laws, rules or other widely accepted principles and uses them to deduce a specific event or structure. Deductive reasoning thus refers to the process of concluding that something must be true because it is a special case of a general principle that is known to be true. Deductive theory also tends to reductionist analysis, as the causes and effects are more finely parsed and tested.

As befits an area of emerging theory, much of the work on clusters is interdisciplinary, normative and inductive in nature. Analysts are looking for meaning from specific observations, hoping that the case method will provide insights or evidence that can be used either to further develop theory or to test some of the basic assumptions and hypotheses derived from theory. Most of the single and many of the smaller comparative case studies are largely historical or contemporary descriptions of the evolution and current scale and scope of the clusters (e.g. Holbrook and Wolfe, 2000, 2002; Wolfe, 2003; Wolf and Lucas, 2004). Many of these works adopt semi-formal or formal quantitative and qualitative survey methods to structure the evidence to support comparative work. The work of the Innovation Systems Research Network in Canada illustrates this. The 12-year research programme has contracts with more than 25 investigators in small, medium and large cities across Canada; given the uneven distribution of primary resources, industrial infrastructure and populations, there was a high degree of heterogeneity in the sample cases. The work in the first phase of the project, in 2000–2005, involved each investigator administering a combined quantitative and qualitative survey of 40 questions with a mix of firms, industry associations and government entities, to investigate the scale and scope of the respondent's organization, their management plans related to research, their relationships with suppliers and customers, their interest in various locational factors (e.g. labour markets, infrastructure, government programmes and educational institutions) and their use and contribution to local social capital. As one might expect from such a diverse set of exemplars and such a wide range of quantitative and qualitative probes, the results were highly variable. The lack of a single unifying hypothesis and method for gathering data to test it led many investigators to deliver inductive, descriptive stories that offered insight into the variety of the origins, structures and outcomes of clusters. Given the idiosyncratic nature of the reports (which were published in a series of six volumes), the lead investigators were only able to derive broad conceptual similarities and differences. A deductive analysis of the hypotheses was not possible. Most other single case studies similarly tell interesting and informative stories but offer no irrefutable evidence of the nature of the systems.

One technological addition some analysts have attempted is to visually map the evolution of specific communities. While technology roadmaps (especially where the engineering or biological links are illustrated) are quite common, they have recently been paired with the institutional structure to illustrate the origins and evolution of technologies and the industrial structures that support them. These maps usually take snapshots of the industrial structure and infrastructure at fixed points in time, comparing and contrasting the changes, especially the spin-outs and mergers. Sometimes these relationships are overlaid on a geographical map to show the role of proximity; often they look like family trees, showing the influence of major industrial actors, government and universities in terms of spin-outs and start-up support (e.g. http://www.doyletechcorp.com/services-trms.htm); in other instances, the relationships determine the location on the visualizations, with closely related firms and institutions clustered in quadrants or sectors of the map (e.g. http://www.globaladvantagebeta.com/innovation-maps/). These maps are designed to show the evolution of a community and sector, with the original enterprise or institution either at the top, left or centre of a visualization with time and linkages displayed (either in a linear hierarchy or in a set of increasingly larger concentric orbits); institutions emerge on the map (sometimes with their shape reflecting their relative size in revenues, employment or investment) at their time of start-up and links show their parents and offspring (e.g. Dobni and Phillips, 2001). While illuminating, the maps are usually a look through the rear-view mirror and seldom offer any clear insights that can be used to make evidence-based policy.

Porter contributed an important extension to the empirical side of clusters by drawing on a number of empirical methods, especially statistical analysis of international market shares and shifting shares over time for key competitive industries (e.g. the Korean experience between 1978 and 1985). While this method offered prima facie evidence of a positive correlation between large or rising market shares and agglomerations of firms in vertically dense industries, many with competitive strategies, two problems remain. First, the cases were not fully compared and analysed – they were individual pieces of evidence that were not combined to test the hypothesis deductively. Secondly, the four factors underlying the diamond of competitiveness were not clearly measured or normalized, with the result that, while the evidence looks compelling, one could make a credible argument that the direction of causality might be reversed – namely, the four factors emerge through the operation of free markets, and are artefacts and not drivers of agglomeration. Others have attempted to address these weaknesses.

One approach pursued by some is to measure and normalize the baseline and differential performance more formally in each of the underlying factors of Porter's diamond model. Padmore and Gibson (1998), for example, developed what they called the GEM analysis – an acronym for groundings, enterprises and markets – which builds on and amplifies the Porter diamond model. The framework focuses on six determinants, grouped into three sections, to focus attention on the critical elements of a cluster and to monitor its development. Groundings, also referred to as supply determinants, are the inputs to the productive process that originate outside the cluster. Enterprises, or structural determinants, are enterprise-based determinants (both public and private enterprises) that influence the efficiency of production of the cluster. Lastly, markets, or demand determinants, are the markets for the firms in the cluster – both final and intermediate demand (including intermediate demand from within the cluster itself). The goal of GEM analysis is to 'reduce the arbitrariness and increase the usefulness of the analytical framework' introduced by past researchers in the area. Once quantifiable elements are identified and measured, they are then normalized to the average of a group of clusters in order to facilitate more in-depth analysis of the variables that drive clusters. To facilitate presentation, Padmore and Gibson (1998) visualized their results as a multidimensional graph, with each cluster ordinally ranked relative to the average of the sample of clusters. The resulting web-like diagram can be used to derive a visual representation of cluster strength and vitality. Moreover, while there is no evidence that anyone has done this yet, the ordinal ratings of a variety of clusters could be used to undertake regression analysis of how these factors affect outcomes (e.g. export market share).

One challenge in cluster analysis is determining the appropriate dependent variable in any statistical or econometric analysis. Porter used export market share and the shifts in these shares as a measure of success (consistent with the product-based view of firm strategy). Two other frequently used statistical measures of industrial concentration are the economic base analysis or location quotient (LQ) and the Herfindahl–Hirschman index (HHI). The LQ, developed by Robert Haig in 1928, is often used to examine the economic base of a region using notions of comparative advantage. More specifically, using employment data, economic activities in a region are measured. The result of the measure is an assessment of the relative shares of total employment in specific sectors or clusters in a region, which can be used to make comparisons and conjectures about the population, income levels and the regional level of expertise or comparative advantage in different industrial activities. Although a relatively rough economic measure of regional industrial concentration and employment, it is still commonly used today. The second measure of industrial concentration, the HHI, named after economists Orris Hierfindahl and Albert Hirschman, has been applied widely to technology management. It is a measure of the size of firms in relation to the industry and an indicator of the amount of competition among them. The major benefit of the HHI in comparison with other concentration indexes, such as the LQ, is that it gives more weight to larger firms.

These and other measures have been used to test for the presence of the clustering phenomenon. Malmberg and Maskell (2002) describe several studies that used industry-level statistics to demonstrate high and rising levels of agglomeration in a variety of industries. These studies included: Krugman's (1991a) analysis of 106 industries across the USA; Enright's (1993) investigation of the forces that shape the agglomeration found by Krugman (1991a); Malmberg and Maskell's (1997) review of industrial agglomeration in Nordic countries over a 20-year period; Isaksen's (1996) use of a location quotient to demonstrate the local labour market agglomeration in Norwegian industries; and Head *et al.*'s (1995) study of the location decisions of 751 US-based, Japanese-owned production facilities. All of these projects were able to demonstrate the existence of agglomeration, which Malmberg and Maskell (2002, p. 436) reported as showing 'support for maintaining the thesis that spatial cluster[ing] at the industry level is a widespread enough phenomenon to justify further study'.

In other instances, empirical research has focused on observing the different types of activity associated with agglomeration. Buenstorf and Fornahl (2006) note a number of studies that focused on the role of spin-offs within emerging clusters, including industries as diverse as the semiconductor industry (Moore and Davis, 2004), the US automobile industry (Klepper, 2004), the US tyre industry (Buenstorf and Klepper, 2005) and the Italian plastics district of Correggio (Patrucco, 2005). Clayman and Holbrook (2004) cast spin-offs as an indicator of a healthy research and innovation sector, especially in the biotechnology and healthcare industries.

Finally, there have been a number of attempts to integrate the descriptive measures above into econometric analysis to identify more specifically causal relationships and their effect on agglomeration. A few illustrative examples give a sense of what is being attempted and accomplished. Krugman and Venables (1996) formulated a theoretical model where agglomeration or clustering is strongly associated with low transport costs but is less likely to occur as transport costs rise. This work led to the important (albeit somewhat counterintuitive) conclusion that the localization of industry is probably associated with globalization and declining barriers to trade. Audretsch and Feldman (1996) provided empirical evidence that clustering increases if new economic knowledge and highly skilled labour are relatively important. In a gravity model study, Leamer and Storper (2001) confirm that production of innovative products has become more agglomerated, while standardized products have become more decentralized. Similarly, Belleflamme *et al.* (2000) showed that decreasing marginal costs resulting from localization economies combined with low transport costs can lead to the forces of agglomeration dominating price competition. In this model, agglomeration is socially optimal but may occur in suboptimal locations or may not happen at all without policy intervention. Finally, Ottaviano and Thisse (2003) concluded their analysis with the observation that agglomeration depends on low-cost market access.

Another group of economists has looked at the role of regional dynamics in creating competitive industries and sectors. Krugman (1998a) and Audretsch (1998) both examined the influence of the Marshall–Arrow–Romer pure external economy, arguing that it explains a significant share of location, especially for innovative firms and sectors. Meanwhile, Porter (1990) and others looked at the role of linkages and market size on production while Zucker *et al.* (1998) looked at the role of specialized labour markets in geographical concentration. Each develops compelling arguments in support of economic agglomerations in clusters.

Stern *et al.* (2001) attempted to model national innovative capacity, combining multiple theories – idea-driven growth (Romer, 1990), national competitive advantage and clusters (Porter, 1990) and bits of national innovation systems (Nelson, 1993). The model measures country-level R&D capacity and productivity, with the number of international patents as an output measure. This work illustrates that broad-based indicators or output measures of national level innovation and productivity pose problems for explaining growth because they fail to account for the wide range of potential interactions of various actors at the regional level. Analyses such as these effectively ignore the potential impact of proximity to key customers, suppliers, educational institutions and rivals at the local level. As a result, the models tend to be biased towards national competency factors, which often are a function of local phenomena.

In spite of the significant effort to develop quantitative and qualitative measures of economic clustering and to use those metrics in economic modelling, the results are generally not overly helpful for policy makers seeking to exploit clusters. The analyses generally support the basic neoclassical policies of open and flexible markets for inputs and outputs but offer only tantalizing hints about which other cluster factors might be important.

CONCLUSIONS

Critics argue that the term cluster is vague and has evolved into a form of rhetoric. From a scholarly perspective, there is limited consensus as to how the term is defined, and there appear to be gaps in the literature, particularly as it relates to identifying and testing causal relationships in order to prescribe appropriate policy measures.

Critics argue that rhetorical excesses and lack of consensus have evolved out of an overuse of the original analytical concept of industrial districts. As noted earlier, Markusen (1999) argued that the cluster literature involves 'fuzzy concepts' that are based on 'scanty evidence', resulting in 'wimpy policy'. According to the OECD (1996), the definition of a cluster 'provides little guidance for narrowing the scope of inquiry in a meaningful way'. Similarly, according to Martin and Sunley (2003), clusters have been 'accepted largely on faith as a valid and meaningful way of thinking about the national economy, as a template or procedure with which to decompose the economy into distinct industrial-geographic groups for the purposes of understanding and promoting competitiveness and innovation'.

This lack of consensus and rigour has led to questions regarding the value of clusters. Chapter 4 contributes to this theoretical, methodological and policy debate with evidence drawn from the agri-food economic agglomeration in Saskatoon, Canada. It offers a practical application of the cluster definition, assesses how economic agglomeration is linked to place-based factors and examines how cluster-based approaches fit with traditional economic and geographical approaches to industrial location theory in the agri-food sector.

4 Policies, Strategies and Metrics Related to Place in Saskatoon

Saskatoon is a spike on the relatively flat economic landscape in the Great Plains region of western North America. As a mid-sized city (265,000 population in 2011), the city provides the higher-order market functions to a population of more than 600,000 that live within 250 km of the city. The city is situated near the centre of Canada's largest crop-producing region – Saskatchewan has about 26 million ha, or about 41% of Canada's total arable land. While the province produces the majority of Canada's exportable grains, oilseeds and pulses and has world-leading market shares of canola, flax, durum, high-quality bread wheat, oats, malting barley and pulses, the region's productive capacity is limited by harsh growing conditions, with on average only between 95 and 125 frost-free growing days, less than 45 cm of moisture annually and extreme seasonality (with daily average temperatures ranging from −20°C in January to 19°C in July, and extremes of −40°C and 40°C). To complicate matters, Saskatoon is more than 500 km from the nearest larger urban centre and more than 1600 km from tidewater.

Since its inception over 100 years ago, Saskatoon has largely been characterized as an urban shopping and service centre. Over the past generation, however, the community was able to bootstrap itself into a national and, for some functions, global hub of agri-food research and innovation. Along the way, Saskatoon has agglomerated a disproportionate share of agri-food research, processing and service functions. Most notably, the city is home to 30% of the nation's agricultural biotech community, with more than 700 scientists working in 30 private, public and academic facilities. By any measure, this is a massively disproportionate share of an internationally important industry, especially given that Saskatoon accounts for less than 1% of the national economy.

This chapter examines the history and structure of this clustering of activity.

THE EMERGENCE OF THE CLUSTER IN SASKATOON

The roots for the current Saskatoon research community extend back to the early 1900s, with the decision to locate the University of Saskatchewan in Saskatoon and its early and sustained focus on agricultural research and teaching (Johnson, 2002). More recently, the federal government's decisions to locate and invest in national agricultural research focused in Saskatoon, beginning in the 1940s, set the stage for early government–university leadership, the attraction of private capital and enterprises, and the evolution of a knowledge-based, agri-food research and development industry. Phillips and Khachatourians (2001) provided a detailed analysis and exposition of the evolution of the canola-related research community in Saskatoon.

Zilberman *et al.* (1997) offered a conceptual framework for understanding the emergence of agricultural biotechnology systems, proposing a five-stage linear development process. Taking this structure and applying it to the Saskatoon centre since the 1950s, one can trace an evolution of the leadership from public laboratories to private corporations and now to producer-led investment pools, all the while mapping out the increasing localization of research, commercialization and production (Table 4.1).

Pattern 1, representing the period between 1950 and 1985, was characterized by public agency leadership in the development of canola, largely through Agriculture Canada (now Agriculture and Agri-food Canada or

Table 4.1. Leadership responsibility for various stages of agri-food product development. (Adapted from Zilberman *et al.*, 1997, p. 6.)

Pattern	Dates	Discovery	Development	Registration	Production	Marketing
1	1950–1985	Public laboratories/ CCC	Public laboratories	Public laboratories	Farmers	Corporations
2	1970–1985	University/ CCC	University	University	Farmers	Corporations
3	1985–date	Public laboratories	Public laboratories	Corporations	Farmers	Corporations
4	1985–date	University	University	Corporations	Farmers	Corporations
5	1985–date	Corporations	Corporations	Corporations	Corporations	Corporations

CCC, Canola Council of Canada.

AAFC) and the National Research Council (NRC). At that point, there was little agglomeration of other effort. Furthermore, there was little collaboration beyond the core public agencies and a few universities, who were quite deliberately scattered across Canada. Pattern 2 began to emerge in the 1970s as farmers exerted more influence through check-off systems (e.g. the Canola Council of Canada or CCC) and industry began to invest in research. Universities (especially Manitoba, Guelph and Alberta for canola and Saskatchewan for small grains and legumes) began to assemble capacity to develop new varieties. This still mirrored pattern 1 with a relatively linear research and development path and limited collaboration. In both cases, after varieties were registered, the seed was marketed to farmers through spot sales, and farm output was almost exclusively purchased and processed or marketed by downstream companies that had little or no interest in the research system.

After 1985, three new patterns emerged. Patterns 3 and 4 replaced patterns 1 and 2, with corporations increasingly setting the objectives of some research programmes (e.g. for canola) and then conclusively taking over responsibility for the registration, multiplication and sale of the seeds to farmers. Downstream, however, the system remained unchanged, with farm output continuing to be marketed by companies for the most part unrelated to the innovation system. This period corresponded with consolidation of research and development of canola in Saskatoon, with private companies and commodity groups investing heavily in research and putting pressure on universities and governments to specialize and create centres of excellence. As a result, the University of Saskatchewan through the Crop Development Centre focused on a few crop species while the federal agencies consolidated their national capacity around oilseed research in Saskatoon (at the AAFC Saskatoon Research Centre and the NRC Plant Biotechnology Institute). This consolidation provided the base for attracting further private research.

Pattern 5 represents a truly new departure from past experience. Private corporations are increasingly assuming leadership and responsibility for more of the stages of product development. This pattern reflects the dominant role of the agro-chemical and global seed companies in the crop-breeding business in recent years. Based on proprietary market assessments, these large firms direct or undertake much of the discovery, development and registration steps, either on their own or through tightly controlled collaborations where they establish the research objectives, pay for the research and assume ownership of the resulting products. In many but not all cases, firms have taken control right into the production and marketing stages, at times supplanting or absorbing traditional seed merchants and marketers. These companies have developed or acquired seed-marketing arms, offering seed to farmers often only through closed-loop contract systems (especially when the seed has some differentiated product attribute, such as modified oil properties), with the corporation buying back the farm product and either marketing or processing it itself. While much of the system has been organized along these lines, other parts remain independent and locally managed and controlled. While one might have concluded a decade ago that much of the regional and national agri-food supply chain would have been reorganized by now, slowing agro-biotechnology investment and development after 1998 has yielded a much wider range of models. In some areas, we can expect that localization might diminish as closely held proprietary marketing channels evolve, while in other stages of development we might expect to see more agglomeration as an embattled set of multinationals seeks new partners to realize scale economies or to gain localized support for research and commercialization.

THE ROLE OF INSTITUTIONS IN LOCAL CLUSTER DEVELOPMENT IN SASKATOON

A survey of participants in the Saskatoon area in the 2001–2004 period offered some evidence that the cluster was maturing according to the Zilberman *et al.* (1997) typology. Sector participants were asked to identify what specific events played an important role in development of the cluster. About two dozen events were cited; six specific events were noted as critical to our understanding of the evolution of the cluster (Table 4.2).

The results suggest that government and the University were the cornerstones of the cluster. The first and most important decision was to establish the University of Saskatchewan in Saskatoon in 1907. The decision to site the University in Saskatoon was hotly debated, but once made, there was almost unanimous agreement that the University should follow the land-grant model developed in the USA, initially focused on offering education, research and extension services to key agricultural parts of the province. The first university training programmes were in agriculture, and the Agricultural College was the second college incorporated in 1912; initially 405 ha of farmland was set aside for the University of Saskatchewan farm for training, research and extension; more land was added over the years, so that the campus is now bounded on two sides by extensive fields. In 1917, the federal government opened an agricultural research centre on campus and began a series of investments at the University. These investments were finally amalgamated in 1959 as the Saskatoon Research Station and have continued to the present day. The third necessary institution was the NRC, the main federal R&D agency in Canada over the past 100 years. While a relative newcomer to the city, arriving only in 1948 with the establishment of the NRC Prairie Regional Laboratory (PRL), it is generally viewed as a key player. Even before the PRL was created, the NRC Industrial Research Assistance Program was developed and is widely credited in Saskatoon as a vital resource that matches technological solutions with firms in need.

Table 4.2. Key events in Saskatoon cluster history. Bold events were judged by survey respondents as necessary events. (From Phillips *et al.*, 2004.)

Period	University	Government	Industry
Before 1950	Created in 1905; agricultural college opened in 1908 (*n* = 5)	F: Ag Canada Dominion Research Lab opened (*n* = 6); NRC-IRAP started (*n* = 3)	
1950–1959	Canola research programmes (*n* = 6)	F: NRC regional laboratory opened (*n* = 10)	
1960–1969		P: SRC created (*n* = 1)	
1970–1979	VIDO created (*n* = 2)	P: Innovation Place opened (*n* = 15)	
1980–1989		F: NRC PBI restructured (*n* = 10); WED policy (*n* = 2); NABI (*n* = 1) P: Provincial S&T policy (*n* = 2)	
1990–1999		F: end of Crow rate and creation of AFIF (*n* = 2) L: SREDA started (*n* = 2); Saskatoon Health District (*n* = 1)	**Ag-West Biotech opened (*n* = 15)**; POS created (*n* = 2); TRLaboratories (*n* = 1); SNN (*n* = 1)
2000–2010	College of Biotechnology in 2000 (*n* = 2); **CLS in 2004 (*n* = 11)**	**F: CLS investments (*n* = 11);** Genome Canada investments (*n* = 1)	20/20 Conference (*n* = 1); BioFutures Conference (*n* = 1); BioProducts (*n* = 1)

F, federal; P, provincial; L, local; see text for other abbreviations.

The latent research capacity of Saskatoon-based institutions was demonstrated with the canola research effort, which began in the late 1940s and culminated in the development of double-zero rapeseed (with low glucosinolates and low erucic acid), which was branded as canola in 1978. The Saskatoon researchers in Agriculture Canada and the NRC and breeders scattered across Canada converted a niche oilseed that has limited food or feed value into a crop that now provides the third largest volume of edible oil in the world. Along the way, the crop has become the major cash crop in western Canada, planted by up to 60,000 farmers annually; the canola supply chain is estimated to contribute more than CAN$13 billion annually to the Canadian economy, according to the CCC.

Local scientists, entrepreneurs and government leaders point to the 1960s and 1970s as a second important period – during these years the province created a provincial research council and opened a research park adjacent to the University (called Innovation Place) and the University and various governments supported the creation of an animal vaccines programme, which has now become the Vaccine and Infectious Diseases Organization and has delivered a range of world firsts in the vaccine field. All are now viewed as important pieces of the local puzzle. The 1980s and 1990s saw a shift in emphasis, as governments moved to provide incentives to firms and farmers to contribute more to the research effort. Many pointed to the restructuring of the NRC operation into the Plant Biotechnology Institute in 1983 as a key event. The effort to refocus national agri-food research was an important factor in the push into biotechnology in Saskatoon. Combined with renewed and expanded public support for private initiative (via Western Economic Diversification and the provincial science strategy) and, perhaps most importantly, the creation of Ag-West Biotech Inc. (a largely publicly funded but privately led industry association, since rebranded as Ag-West Bio), governments set the stage for industry to engage in and at times take the lead in setting research priorities. A range of multinational corporations from the USA and Europe subsequently made local investments in both collaborative research programmes and in proprietary R&D facilities. While Saskatoon's cluster began to make the transition to stage 5 in the Zilberman *et al.* (1997) typology, it is important to remember that the industry and its leaders remained largely dependent on public funding. The collective institutions, such as Ag-West Biotech, the public laboratories and many of the private firms all receive significant public funds – so, although leadership shifted, the funding of the initiative remains dependent on public support.

In the same survey, business, university and policy leaders were asked who the most influential people were in the innovation cluster. A total of 64 people were recognized as leaders throughout the interviewing process. While there is significant potential for faulty memories or revisionist history, the results in Table 4.3 drawn from the entire sample suggested that there were classes of leaders at various points over the life cycle of the cluster. Given that this question was asked late in the development period, there was probably a bias towards acknowledging private-sector leaders, such as the two most recent Presidents of Ag-West Biotech, in contrast to earlier university and government leaders. Nevertheless, the data suggested that certain classes of leader had more influence than others. Over most of the sample, university presidents and key scientists and administrators from the two federal laboratories – AAFC and the NRC Plant Biotechnology Institute (NRC-PBI) – were unambiguously viewed as key. Individuals from firms, the University faculty and virtually all political, programme or policy actors were for the most part acknowledged but given little credit for the success of the cluster, which would suggest that, while traded interdependencies may be a necessary condition for the cluster, they were not as central as Porter's (1990) theories would suggest.

Table 4.3. Key business, community and government leaders in the Saskatoon cluster. (From Phillips *et al.*, 2004.)

	Individuals mentioned		Citations	
Sector and institution	No.	% of total	No.	% of total
Industry	9	14.1	51	32.5
Ag-West Biotech	2	3.1	31	19.8
Private firms	6	9.4	19	12.1
Chamber of Commerce	1	1.6	1	0.6
University	16	25.0	27	17.2
Administration	9	14.1	16	10.2
Faculty	5	7.8	8	5.1
CLS	2	3.1	3	1.9
Federal government	18	28.1	42	26.8
Politicians	2	3.1	2	1.3
Bureaucrats	3	4.7	3	1.9
AAFC	5	7.8	11	7.0
NRC-PBI	5	7.8	22	14.0
NRC-IRAP	3	4.7	4	2.7
Provincial government	12	18.8	17	10.8
Politicians	5	7.8	8	5.1
Bureaucrats	5	7.8	5	3.2
Innovation Place	2	3.1	4	2.6
Municipal politicians and administrators	2	3.1	13	8.3
Other (Not specified elsewhere)	7	10.9	7	4.5
Total	64	100	157	100

UNIVERSITIES AS ENGINES OF GROWTH

The history of the Saskatoon cluster suggests that one might realistically look at the University as a long-term and fundamental contributor. There is ample evidence from elsewhere that a regional university can directly and indirectly drive a technology cluster. Niosi and Bas (2001) stated that innovation can be spurred in emerging technologies and industrial clusters indirectly through decentralized and horizontal policies to support the creation of both government laboratories and research universities. Universities are most often credited with generating know-why, basic scientific knowledge, being a source of skilled labour, acting as a magnet (directly and indirectly) for research stars and businesses, and being a conduit for both local and non-local knowledge exchanges (i.e. publications).

The traditional role of a university is to generate and diffuse basic or explorative (know-why) knowledge as well as to generate a skilled academic and technical labour force. However, these traditional roles are evolving. As Cooke (2002) argued, a strong local science base needs to be complemented by a thick entrepreneurial culture not only within the regional business community but within the academic community as well. Cooke (2002) suggested that '... the science base is a magnet, even if only indirectly... for biotechnology business', which is due in large part to the technology's innate, strong scientific knowledge content. This is supported by previous research, which found that the existence of a diversified, mainly academic, knowledge base is a prerequisite for successful localized innovative activity in biotechnology (Zeller, 2001). For instance, Vancouver has at times had many of Canada's fastest-growing firms, which some attribute to the presence and activities of the University of British Columbia (J. Niosi, personal communication, 5 December 2003). The University of British Columbia has generated three times as many biotechnology-related patents as the University of Toronto. Niosi also suggested that universities are a foundational element of the 'virtuous cycle' embedded within the cluster phenomenon – star scientists become entrepreneurs and work to spin off commercial ventures. This virtuous cycle is explicitly linked with university-based capacity.

While universities appear to play a key role in clusters, there is a lack of consensus as to how universities underpin innovative activities in the context of a globalized learning economy (Lundvall and Borrás, 1998). As academia no longer holds a monopoly on learning and knowledge generation, it is becoming accepted that effective universities are not 'ivory towers' isolated from other actors within the region but rather function as partners and competitors with both local and non-local actors.

The University of Saskatchewan and the Innovation Place Research Park located on university land adjacent to the academic campus work closely together on a variety of projects and collaborations. The two play host to several key public institutions, including the NRC-PBI, the federal AAFC Research Centre, the Saskatchewan Research Council (SRC), and more than 50 companies, commodity groups and other important employers and service providers. This provides students, faculty and researchers with the opportunity to work in world-class facilities with state-of-the-art equipment and, most importantly, in collaboration with highly skilled research personnel.

The University of Saskatchewan is by far the largest employer of life-science-based faculty and research staff. In 2004, the University reported that it employed 422 faculty and sessional lecturers in five life-science colleges and 370 full-time-equivalent research staff in the broad subject area. While the single largest number of faculty and researchers were in the medical area, the two most research-intensive colleges were Agriculture and Veterinary Medicine, which combined together employed almost 145 faculty and lecturers and 210 research staff. In total, almost half of the University faculty, teaching and research resources are focused on life-science education and research. This faculty has access to 51,185 m^2 of laboratory space in 1269 rooms on campus.

The university provides two key measurable outputs: (i) educated and skilled graduates; and (ii) intellectual property. In 2002, the University had an estimated 344 graduate students enrolled in research programmes in the five life-science colleges. Almost two-thirds of those students were in the area of plant and animal research in the Agriculture and Veterinary Medicine colleges.

A survey undertaken in the Saskatoon agro-biotechnology community in 1998 illustrated the critical role of the local education system (Table 4.4). All but two of the employees surveyed with a technical diploma were trained in Canada and 82% of these got their training from the Saskatchewan Institute of Applied Science and Technology, mostly through the 2-year Biotechnology Program (now BioScience Technology Program) in Saskatoon. Between 1969 and 1998, the programme produced 335 graduates (SREDA, 1998), and employment surveys of graduates showed that 88% were employed in their field within 6 months of graduation. With the acceleration of activity in Saskatoon in the late 1990s, the enrolment rate in the programme rose to approximately 15 per year, up from under ten per year in the 1980s. The same pattern was seen among employees with undergraduate training. All but 22 of 85 of the respondents with a bachelor's degree were trained in Canada. There has been a bit more mobility at this level of training, however, with only 66% of the respondents getting their training at the University of Saskatchewan. The Saskatoon labour market draws from Ontario, Alberta and Manitoba extensively. The labour market becomes significantly more mobile at the graduate level. Less than half the employees with master's degrees and only about one-quarter of the employees with doctorates were trained in Saskatchewan.

Table 4.4. Employees in the Saskatoon agri-food research community, by degree, in 1998. (From Phillips and Khachatourians, 2001.)

	Technical diploma	Undergraduate degrees	Master's degrees	PhD degrees
Total no.	88	235	67	73
Saskatchewan (% of total)	82	66	43	27
Canada (% of total)	98	91	84	64
USA (% of total)	1	2	6	10
Europe (% of total)	0	4	7	18
Other (% of total)	1	3	3	8

In the early part of the last decade, collaboration with the provincial government led to the founding of a *virtual* College of Biotechnology at the University of Saskatchewan. In operation from 2000 to 2007, the virtual College of Biotechnology offered nine undergraduate degree courses, at its peak in 2003–2004 to more than 300 students, that provided students with the opportunity to learn about the scientific, social, legal, commercial, economic and ethical aspects of biotechnology. The college ultimately saw more than 250 students graduate, many of whom found work in the local research and industrial community. The university also introduced a 1-year taught MBA in Biotechnology Management, which operated for a number of years before being repurposed.

Equally important for the community is the flow of intellectual property from the University. In 1990, the University of Saskatchewan Technologies Inc. (UST) was established as the technology commercialization arm of the University (now called the University of Saskatchewan Industry Liaison Office). The office aids researchers in developing optimal strategies for intellectual property protection and offers help to guide them through the commercialization process for their innovations, including market development. Since then, there have been more than 25 spin-offs from the University. In terms of intellectual property created through the University, a patent search on the Canadian Intellectual Property Organization website found that 60 patents were filed by the University of Saskatchewan before 2000 and another 70 filed between 2000 and 2012. A related patent search of the US Patent and Trademark Office patent database revealed 37 patents issues to the University in the 2001–2012 period, mostly for life-science inventions. An analysis by Niosi and Bas (2001) showed that in the 1989–1999 period, Saskatoon ranked fourth in Canada for patent issues (behind Montreal, Toronto and Edmonton, or Vancouver, depending on the measure). Furthermore, the biotechnology-related patent activity was proportionately much higher in Saskatoon (67% of all university patents) relative to other universities in Canada (average 16% for all universities).

PRIVATE ACTORS IN THE SASKATOON CLUSTER

There is some debate about the role and appropriate structure of firms in a cluster – is it better to have small, entrepreneurial firms or large multinationals? Saskatoon has tested both models.

One insight from scholars looking at industrial structure (e.g. Enright, 2000; Dunning, 2001; Rugman and Verbeke, 2001) is that most clusters have MNEs embedded as leading participants – an oft-cited agri-food example is Monsanto in the St Louis Missouri BioBelt. Rugman and D'Cruz (2000) argued that MNEs act as flagships to lead, direct, coordinate and manage strategic value-added activities of collaborative firms in the combined business and social networks. They explored the extent to which the largest MNEs are regionally based. According to this research, only six of the 20 MNEs with the highest internationalization of their activities (defined by revenues and global presence) can be considered candidates for the status of MNEs with a global strategy; the remainder are largely home based. Alternatively, Gassmann and Gaso (2004) looked at the role of MNEs as brokers of knowledge in clusters and proposed that they operated like listening posts in many communities.

While MNEs have been and are present in the Saskatoon cluster and at times have been important actors, they have not been the defining institutions in this cluster. The size and composition of the Saskatoon cluster is dynamic. New firms and programmes emerge at the same time as obsolete programmes and projects disappear and firms migrate to other sites. In 2002–2004, private actors (i.e. those involved in either development or commercialization activities) represented only about 33% of the local population (n = 110). The majority of private actors were found to have fewer than 50 employees. Among this group were a number of MNEs that maintained regional offices and undertook sales and research – performing the knowledge-broker and listening-post functions identified by Gassmann and Gaso (2004). Only three private actors had more than 50 employees. All three had developed out of local competencies or discoveries (Becker Underwood, Bioriginal and Philom Bios). In addition to being relatively small, the majority of private actors were young (73% of private actors were established after 1990).

To get a better understanding of the role of locational factors in firm decisions, 22 of the 37 firms were interviewed. Eleven of the respondents reported that their enterprise was founded in Saskatoon because of a recognized gap in the local market. Five of the actors interviewed created a new office to represent larger, non-resident corporate interests and six companies were spin-offs, either from the University or from other corporate organizations. Ultimately, firms are created by and for people. When we delved deeper into the logic for the firms locating in Saskatoon, the dominant response related to people factors – Saskatoon was the home of the founder for about half of the firms.

Nevertheless, firms reported variables encompassed in the broad competitiveness model of clusters. In order of declining importance, firms reported that they gained value from an ample supply of skilled workers, appropriate market signals, specialized infrastructure and institutions (e.g. the University and the Canadian Light Source (CLS)), the agglomeration of other firms and actors and a set of related cost efficiencies and specific research competencies and capacities (e.g. canola and the local base of land for field trials).

In short, unprompted, the firms themselves acknowledged the four traded interdependencies of cluster theory: (i) demand conditions; (ii) related supporting industries; (iii) factor conditions; and (iv) firm strategy, structure and rivalry.

Demand conditions

Marshall, Schumpeter and Porter, among others, have all pointed to the importance of local markets, especially their ability to convey meaningful and strategic advice on the nature of the quantities and qualities of products that markets demand now and into the future. Ultimately, consumer demands that compel firms to improve their competitiveness through innovative products are critical to the long-term viability of a cluster.

Research conducted in 2003–2004 on the dynamics of the biotechnology cluster in Saskatoon revealed that almost two-thirds of the companies interviewed indicated that, once they were producing a product, the value of co-location diminished. While relationships with customers are important, co-location is not critical – 82% stated that relocation to be near key customers would not be considered. There was more diversity of opinion about whether it was important to be located near their suppliers. About 45% reported relying solely on non-local suppliers and another 32% reported that they used both local and non-local suppliers; more than half reported that it was not important to be close to suppliers. Lastly, it was interesting to note who the companies viewed as competitors. Approximately half of the actors recognized their competitors as global, while 41% reported only local competition. In contrast to ambivalence about co-location with suppliers or buyers, more than three-quarters of firms reported that there was real value in being located near competitors. One could conclude that many firms at the R&D and early commercialization stage see co-location as improving their odds of competing.

Moving downstream from the research stage, there are vital forward linkages that increase Saskatoon's attractiveness as a research site. This is actually a two-step process, as regulators must approve new genetically modified (GM) crop varieties and consumers must buy and consume the products of biotechnology. More than one-fifth of all companies surveyed in 2002 (Phillips, 2002), representing most of the larger breeding operations, indicated that rapid access to a receptive seed market was critical to their decision to locate in Canada. They confirmed that undertaking the research and commercializing the resulting varieties under the same regulatory system was a key feature in their location decisions. In the first instance, the regulatory system determines how quickly farmers can and will adopt new varieties. The Canadian regulatory system is generally viewed as somewhat slower than the US system, but is clearly faster and more predictable than those in Europe and Australia, where transgenic crops have yet to be fully adopted. Beyond the regulatory decision, farm programmes can partly determine the pace of adoption of new varieties. In Canada, new varieties of canola are automatically eligible for crop insurance and other stabilization assistance, without limits on acreage seeded or location. In the EU, in particular, increases in acreage of a new crop are often restricted by domestic or international agreements. Following regulatory decisions, farmers have to assess, adapt and adopt varieties to bring the technologies to fruition. They rely on signals from suppliers, the wholesale system they sell into, and the pricing and conditions of access to new-trait crops. The evidence is compelling that the more than 100,000 farmers in western Canada are rapid adopters of new varieties – herbicide-tolerant canola, for instance, reached more than 90% adoption in less than 5 years among a group of more than 60,000 producers (Gusta *et al.*, 2011).

Extensive farm service networks support the large, receptive and relatively sophisticated farm market for new seeds in the prairies. In addition, a far-reaching network of not-for-profit producer and industry organizations such as the CCC and the various provincial growers associations ensure the rapid and efficient adoption of the technology.

These producers then apparently seamlessly connect to the largely private and competitive wholesale grain handling and distribution system, enabling western Canadian producers to dominate in a number of important high-value crops where the underlying traits were developed in and around Saskatoon.

Related supporting industries

The theory of clusters lays out fairly clearly the importance of spatial proximity of upstream or downstream industries in a firm's product value chain in facilitating the exchange of ideas and the resulting innovations.

Saskatoon-based firms, however, were unlikely to refer to formal collaboration and cooperation with other firms as a major part of their business strategy. Only two firms cited collaborations with other firms as a strategic advantage and, in those cases, they were referring to a parent or sister firm. While many firms reported that they collaborated for components of their innovation process, such as product testing, in many cases this was a transaction that was indistinguishable from merely purchasing the services of another firm. Tellingly, when asked to describe a product, process or service developed in the last 3 years that was representative of their innovativeness – a question that most firms, understandably, used to exhibit their most impressive innovation – only one firm talked about a product that was developed in close collaboration with another firm or organization. Otherwise, each firm described a product created solely by internal R&D, although often with inspiration from outside sources.

Economic theory indicates that firms locating where there are extensive backward linkages into supporting industries and forward linkages into the market can realize economies of scope. In Saskatoon, however, apart from publicly provided infrastructure, there is limited evidence of critical mass of other specialized industry that strengthens the bond between the canola research industry and the location.

Private venture financing is limited, with no access to a local stock market (Calgary is the nearest market) and with only limited venture capital lending. The public sector fills the gap. Two financial institutions – the Royal Bank in Saskatoon and the Canadian Imperial Bank of Commerce (CIBC) in Winnipeg, both in partnership with Western Economic Diversification – have operated specific knowledge-based lending facilities, while there is limited public lending capacity with Western Economic Diversification, the Business Development Bank of Canada, the Saskatchewan Opportunities Corporation and the Agri-food Equity Fund, all having offices in Innovation Place or Saskatoon. While small and varied, the operation of the various federal, provincial, regional and local tax and fiscal programmes has sharply reduced the cost of locally conducted research and development to the point where, on aggregate, much of the private innovation enters the market tax free (Phillips and Khachatourians, 2001). Publicly provided infrastructure, company-specific industrial incentives, federal and provincial matching research funds, research in the public laboratories, public financing of university research and a plethora of grants and tax incentives combine to underwrite virtually all of the out-of-pocket costs for private research in Canada. Furthermore, the privileged tax status of western Canadian farmers ensures that producers retain most of the benefits from early adoption. Hence, in spite of weak private capital markets, the public system has effectively made the biotechnology sector tax free.

Although the financial, accounting and legal communities have been restructuring recently to service the growing private research effort in Saskatoon, surveys suggest that only limited accommodation had been made to support and service the developing biotechnology industry. All of the multinational firms undertaking research in Saskatoon have indicated in discussions that they do their banking corporately, with much of the service coming from their Canadian or global headquarters, which are always located in another city or province. Meanwhile, the Chartered Accountants of Saskatchewan, the Certified Management Accountants of Saskatchewan and the Certified General Accountants of Saskatchewan, indicate few listings for practices that deal specifically in the area of biotechnology. Surveys of biotechnology companies undertaken similarly reveal that multinational firms generally use accounting services purchased through their Canadian head offices, while smaller firms use local accountants and auditors. Few accounting practices market themselves as having a speciality in accounting for knowledge-based enterprises.

A third key service required by knowledge-based firms is legal support for protecting intellectual property that may have commercial value. A key limiting factor is that, until recently, there were no resident patent agents in Saskatoon. While there are now a few available locally, most of the patent work continues to be done in Calgary, Edmonton, Winnipeg or Ottawa. Moreover, few multipartner law practices operating in Saskatoon have ongoing and sustained capacity in the area of intellectual property rights. The focus of the local legal community is mostly on Canadian law, which is a challenge as innovators often wish to protect most of the innovations internationally. In short, firms are being serviced, but the presence of many multinationals, which buy these services internationally, combined with the restricted local capacity, limits the potential traded benefits that could accrue to the industry.

Factor conditions

Most studies of clusters look at a range of local factor conditions, such as specialized factors of production, skilled labour, capital and infrastructure, and their role in clustering.

A 1998 survey of about 30 firms undertaking R&D related to canola globally but engaged in the research effort in Saskatoon, either directly or indirectly, investigated the factors influencing their decision to locate their research efforts (Phillips and Khachatourians, 2001). Half of all respondents, representing the majority of private companies responding, acknowledged the importance of proximity to either collaborators or competitors, especially the NRC and AAFC and key research universities.

In 2008, we surveyed 24 firms from two recognized and established innovation clusters in Saskatoon – predominantly from the mining and agri-food sectors. Generally, these firms credited their competitive advantage to superior innovation; more than half expressed a belief that this was due to their own efforts or to their international connections or networks of alliances and partnerships. There was a marked difference between the clusters. Almost all of the firms involved in agricultural and biotechnology research believed that the quality of their research and development was a competitive advantage that was derived from their relationships with contract service providers, customers and large-scale projects involving a number of major specialized research facilities, such as the University, the SRC, NRC-PBI, the POS Pilot Plant, AAFC, the local Industrial Research Assistance Program (IRAP) office, the Innovation Place Bioresource Centre and the Vaccine and Infectious Disease Organization (VIDO). When asked to describe the motivation underlying their collaborations, the most common answer was the need for efficiency and to cut costs; other answers revolved around gaining access to knowledge, innovation and expertise in order to stay at the cutting edge of advancements in science and technology. This community-wide result correlates with the results of earlier work specifically focused on exploring the canola sector (Phillips and Khachatourians, 2001) and, more broadly, with work that identified as important many of the same institutions and key individuals within those institutions.

The local labour market is also credited as a key asset in the development of the community. The gradual agglomeration of agri-food research (especially for canola) in Saskatoon and Canada suggests that there should be some evidence of a specialty labour market evolving. More than one-quarter of companies responding to a canola industry survey, and the clear majority of private companies, indicated that access to a deep, local, skilled labour pool was important. A survey of canola research employees in Saskatoon undertaken in the summer of 1998 received 390 responses (from 1000 surveys distributed), representing 169 person-years of canola-related employment in 1998, compared with the 248 person-years of related employment in Saskatchewan identified in the company survey. When the degree-granting institutions were examined, the dynamics of the industry become pronounced. As noted above, the higher the qualifications, the greater the mobility and cross-national movement of employees. At the PhD level, more than 35% of the workers were trained offshore in Europe, the USA and other countries. Nevertheless, it is important to note that the single largest source of employees at all levels was the local university in Saskatoon.

The data show that, although the local labour market is able to supply much of the labour required, some of the more highly skilled employees need to be recruited from elsewhere. Hall (1999, p. 498) emphasizes the role of mobility in innovation – 'probably no city has ever been creative without continuous renewal of the creative bloodstream'. In the 1998 survey, all employees in Saskatoon asked what features of the job and community affected their willingness to move to or from the city. Somewhat surprising, given conventional wisdom in the industry that quality-of-life concerns would dominate, the thickness of the labour market was the key consideration mentioned by respondents. All employees with graduate degrees that responded to the question ranked proximity to other companies or agencies that could hire them as a top five consideration and 87% of the respondents put it as the most important consideration. The second most important feature was the type of work in the job, another feature of a thick labour market. Salary and benefits came third, followed closely by career prospects. Almost all the other factors, either related to the job or related to the community, were ranked well below these four factors (Phillips and Khachatourians, 2001).

Zucker *et al.* (1998) offered an approach to determine more explicitly how the labour market can contribute to agglomeration. Their study examined the role of human capital in the birth of US biotechnology enterprises by looking for causal links between the location of research stars and the creation of new firms. They defined stars as scientists who had discovered 40 or more genetic sequences or scientists who had written 20 or more articles on genetic sequence discoveries. They concluded that the presence of active stars in a region was strongly positively correlated with the start up of new ventures, stating that 'at least for this high-technology industry, the growth and location of intellectual human capital was the principal determinant of the growth and location of the industry itself'. Phillips (2002) undertook a comparable analysis of research stars in the canola sector, defined as those who had published at least 20 articles (borderline stars published 15–19 articles). This study found 69 individual scientists worldwide fitted the criteria: about 45% in Canada, 45% in Europe and 10% in Japan. Approximately 63% of the borderline stars were in Canada and the rest were scattered in Europe, Japan, the USA and Australia. In total, the 69 stars and near stars, which represented just less than 1% of all scientists working on canola, produced 1523 articles, or about 31% of all articles produced over the period (Zucker *et al.*, 1998, found that the stars in their study represented 0.75% of all scientists but 17% of all articles). The largest single geographical concentration of stars and near stars in the world was in Saskatoon, where 16% of the scientists lived and worked. The stars and near stars were then assessed by their citation rates, showing that Saskatoon had six out of 40, or 15% of the total and about one-third of all the Canadian stars and near stars.

Moving beyond the labour market, there is some evidence that the Saskatoon biotechnology cluster generates specific factors that offer both tangible and intangible benefits. Procyshyn (2004) used the results from the 2003–2004 surveys in the Saskatoon biotechnology cluster to examine the role of eight central actors in the biotechnology cluster. Each central actor was asked to identify their relationships with each of 94 other institutions in Saskatoon, noting whether they were engaged in exchanges related to R&D, fee-for-service work, financial capital, human capital or contextual knowledge. A social network map was created and density measures for the entire network and each of the five functional networks were developed. On average, there were 89 linkages per core institution, spread across the five functions analysed. The overall network density across all functions was 15% (712 linkages identified relative to an absolute possibility of 8930). As one would expect, the density varied across function, from a low of 1.4% for financial exchanges and 2.3% for R&D links (which is consistent with a predominantly pre-commercial research-based community) to a high of 7.6% for networking, which suggests a highly linked community for industrial policy and promotional efforts but relatively weak networks for financial intermediation.

Another important feature in cluster theory is the role of highly specialized infrastructure. Saskatoon has developed some innovative and unique hard and soft infrastructure targeted on nurturing competitive and innovative agri-food products. Hard infrastructure is particularly important to Saskatoon's economic success. Investment in gross fixed capital has equalled about 21% of Saskatchewan's GDP in the past decade (ranging from 19 to 23% on an annual basis). Saskatoon's traditional share of provincial investment has always been higher than its proportion of population, and evidence suggests that this share rose sharply in recent decades. Federal, provincial, municipal and industrial investors have responded to economic opportunities and demographic pressures in the city in recent years. In the past decade alone, more than CAN$1 billion of new investment has been directed to the University and related research facilities and major new capital has flowed into oil, heavy oil, potash, uranium, gold and diamonds, all headquartered and served from Saskatoon. Despite the 2008–2010 global recession, commodity prices stayed relatively robust and the Saskatoon economy remained buoyant and a prime investment target.

Since its inception in 1997, the Canada Foundation for Innovation (CFI) has been a major investor in Saskatoon. The two single largest contemporary investments in economic development in Saskatoon are the CLS and the International Vaccine Centre (InterVac). Approved with six beamlines at an initial capital cost of CAN$173 million, the CLS, a third-generation synchrotron, is the largest science project in Canadian history and was the product of an unprecedented collaboration of federal, provincial and municipal governments and agencies, universities from across the country and industry. As of 2011, eight additional beamlines were funded and developed, each at an additional cost of at least CAN$10 million. In 2008, VIDO at the University of Saskatchewan was commissioned to build InterVac, a CAN$140 million vaccine research and development facility, which will be the largest Containment Level 3 facility in western Canada developing vaccines for humans and animals against emerging or persistent diseases. The two projects together account for more than CAN$400 million of investment in recent years. The tag line for the CLS – 'field of beams' – evokes the hopes of Ray Kinsella in the 1989 hit movie *Field of Dreams* that 'if we build it, they will come'. The notion that infrastructure imbues some irresistible draw for science and investment is not unique to the CLS – the supply–push notion of investment as a primer for development permeates the literature and myths of economic development. While much of the identifiable innovation in Saskatoon industry is linked to one of these hard-infrastructure specialized investments, not all of the facilities can claim credit for significant innovative success yet.

Meanwhile, soft infrastructure has been a significant focus of government policy with respect to innovation in recent years. One long-term and significant source of targeted funding that nurtures research networks has been through the commodity check-offs, which are facilitated by federal and provincial legislation. Producers pay small levies on crops and livestock produced and marketed and use that to leverage university and industrial research and development. While the individual levies are small, the aggregate funds available in Saskatchewan for research have exceeded more than CAN$100 million over the past 20 years. Much of this has been coordinated and used by groups such as the Crop Development Centre, POS Pilot Plant and various federal laboratories, all located at the University. Meanwhile governments have also directly put significant money into soft infrastructure. The province has contributed by developing groups such as Ag-West Biotech, a non-profit Saskatchewan corporation funded by the government but with an independent Board of Directors, which is mandated to 'initiate, promote and support the growth of Saskatchewan's agricultural biotechnology industries' and 'the commercialization of related food and non-food technologies'. The city, meanwhile, created the Saskatoon Regional Economic Development Authority (SREDA), one of the first independent economic development agencies in Canada, to coordinate and lead economic and community development. Nationally, funding agencies put money in networked science and research, including through the Networks of Centres of Excellence, Genome Canada (GC) and various new tri-council-funded networking initiatives, all of which Saskatoon has worked to use to generate new activity in the city.

Most of this investment in both hard and soft infrastructure is pushing the local economy into further specialization and away from diversity. The laboratories and facilities are usually single purposed, and the grants and capital pools are increasingly focused on clearly delimited outcomes. While this undoubtedly has created some local buzz – respondents to various surveys report the important role of the University, NRC-PBI, AAFC and Innovation Place – the newer investments are highly specific. There is some recognition that these new investments may need some special soft infrastructure to optimize their benefits – the University, for example, has proactively sought to attract research-intensive scholars to take up Canada Research Chairs on campus to use the CLS, while the University business office has aggressively pushed for more work with downstream users. The jury is out as to how much soft infrastructure is needed to ensure that hard infrastructure investment will be able to contribute fully to the creative and innovative community.

Firm strategy

Karwandy (2008) took on the task of assessing and empirically testing the hypothesis that firm strategy, structure and rivalry – i.e. the dynamic conditions that compel firms to innovate – are important, measurable variables. Using the competitiveness theory of clusters articulated in Chapter 3, he set out to test the proposition that proximity fosters sustained trade linkages that shorten communication lines, reduce costs (e.g. transaction, search, negotiation and enforcement), intensify rivalry and deliver more and better intelligence on competitors and markets. The platform for these effects is the exchange-based relationships among organizations and their suppliers, distributors, competitors and customers. Hence, the value chain focus of the product-based view of management best fits with the performance objectives that should flow from clusters. Specifically, organizations are assumed to pursue dominance over their competitors by managing their value chain relationships so as to maximize the margin (the gap between production costs and the customer's willingness to pay). Value chain relationships contribute to this objective by removing uncertainty surrounding input costs and sales opportunities, and by providing information about changes in the competitive environment. Thus, Karwandy proposed that the competitiveness perspective on clustering most conforms to the product-based view of management. The product-based view of management posits that competitive dominance can be represented by four key measures: scale of corporate revenues, the presence of growth, and the importance of international and domestic sales. Karwandy then developed three variables that reflect traded interdependencies of the cluster, including the local share of customers, suppliers and competitors.

A sample of 92 firms and agents in the biotechnology and new media sectors in Vancouver and Saskatoon were examined. The data for seven variables (three representing measures for traded interdependency and four output measures that firms pursing product-based strategies would use) were drawn from semi-structured interviews conducted in 2002–2003 that were taped, transcribed and then coded by a single investigator (to ensure consistency in interpretation). The behavioural or causal strategy variables were then compared with the product-based success measures. Due to small sample sizes, non-parametric tests were used, as these make fewer assumptions about the distribution of the sample population and can be used to analyse nominal and ordinal scale data. The Kruskal–Wallis test was calculated and a χ^2 test was used to determine significance. The analysis was conducted in two stages. First, all the tests were run against a subset of 49 biotechnology respondents. Following this, the new media respondents were added to the analysis to create a total sample number of 92. By adding the new media respondents, one can expect to observe the model's sensitivity to sample size as well as the sectoral orientation.

In the smaller sample of 49 biotechnology firms in the two sectors, only three pairings out of 12 tests produced statistically significant relationships. The tests offer some support that, for biotechnology firms, the local presence of key customers is weakly related to firm growth while the local presence of key competitors is positively correlated (albeit not strongly) with the share of both a firm's domestic and international sales. The Kruskal–Wallis tests returned non-significant results for all but two pairings. The relationship between the presence of local competitors and international market share was weakly significant – at the 90% confidence level, it explained only 20% of the variability in the dependent variable. The other significant relationship (at the 95% confidence level) was between local suppliers and gross sales, but once again the relationship explained only 15% of the variability in revenue. These findings largely conform to the results of the canola industry survey (Phillips and Khachatourians, 2001), which identified the importance of competitors and collaborators to firms in this sector but offer only limited support for the assertion that place-based, traded interdependencies accessed in clusters fully explain superior economic performance among firms pursuing product-based strategies. In short, there may be value in the theory, but it is not overwhelmingly compelling.

THE LOCAL–GLOBAL NEXUS IN SASKATOON

In practice, standing back from either the firm or the region and looking instead at the sources and uses of knowledge in the system gives a significantly different perspective. Drawing on evidence from the agricultural research world through the eyes of the Saskatoon canola sector highlights the importance of supply chains that transcend any firm or region. No one firm or region drives the activity – multiple actors jointly lead it. Undoubtedly, parts of the innovative effort have the potential to become visibly linked to firms or regions, but the balance between local and global links is important to understand. In essence, clusters, especially innovative ones, reflect many of the characteristics of the classical trade *entrepôt*, where many of the key inputs are imported, value is added in a local milieu and then semi-finished or finished outputs are exported for further processing or consumption (Endacott, 1964).

The Saskatoon-centred, canola-based biotechnology cluster exhibits many of these attributes, with the bulk of the basic research and many of the proprietary technologies developed elsewhere being imported, assembled into new crop varieties for commercial release first in western Canada and then exported on a largely tax-free basis as germplasm or intermediate product to global markets (Phillips, 2002). Consistent with both the economic and cluster theories, canola research agglomerated in and around Saskatoon, which poses significant implications for researchers, markets and public policies.

The dimensions of the Saskatoon innovation *entrepôt* can best be analysed by looking at its relative role in creating knowledge, using knowledge and commercializing new products. One might conclude that Canada is the main canola innovator based on its record as the lead innovator and early adopter of all the new traits over the past 40 years. But a significant share of the applied research to develop the processes used to create those varieties has been done in other countries and much of the applications-based research (e.g. uses for new oils) is happening elsewhere. This suggests that Canada operates in a niche in this global knowledge-based industry – as an *entrepôt* undertaking and assembling the know-why, know-how and know-who of varietal breeding and primary production – but that the bulk of the activities upstream and downstream of that stage in the production system are now and may continue to be done elsewhere. This will be discussed further in Chapter 6.

CONCLUSIONS

The theory outlined in Chapter 3 offers some hard and fast hypotheses that many regions have adopted as policy prescriptions, often without testing the validity of the underlying assumptions. This chapter has presented a range of studies in Saskatoon that suggest greater care and attention is needed to the details. While many are tempted to accept that success is validation of the theory, the research presented in this chapter suggests that one must carefully investigate the causal pathways rather than simply accept notional correlations.

The surveys and studies in Saskatoon have shown only weak correlations between the assumed traded interdependencies that underpin clusters and for the most part challenge the assumption that competitive, strategic firms will necessarily benefit from co-location to access traded interdependencies.

The great risk of the place-based cluster theory is that policy advocates at times take the causality as given and then seek to build local capacity based on a policy of autarky, with preferential tax, subsidy, zoning and procurement policies building local traded capacity at the expense of more nuanced structures. Landry (2004) showed in the context of a study of manufacturing in Quebec City that, while there is no definitive tipping point, there is evidence that, as the local density of cluster linkages increases, especially relative to links outside the cluster, firm performance deteriorates.

The theoretical and empirical exercise of unpacking the local cluster into a knowledge system (tracking largely traded interdependencies) revealed that Saskatoon, and by extension probably many other innovative clusters, exhibit a highly complex mix of local and global linkages that appear to contribute to success. The *entrepôt* model of development is counterintuitive and unpopular with many scholars and practitioners but would appear to be a useful rubric to use to avoid missing critical causal factors.

5

Local Links and Global Pipelines: Innovation Systems Theories and Models

'Space, the final frontier', a common phrase in popular culture coined by Gene Roddenberry in *Star Trek*, brings to mind the unprecedented and brave pursuit of the new and undiscovered. It evokes the same intent of Vannevar Bush's 1945 report to the US President – entitled *Science, The Endless Frontier* – which concluded that basic research and the organizational system that propels research, development and commercialization was 'the pacemaker of technological progress'. The space and science frontiers are presented in some ways as abstract, nebulous and something to conquer.

Although of another time, Marshall (1890) suggested that the effort to push out the frontier can more effectively be done through economic agglomerations – proximity would lead to 'mysteries in the air', the tacit externalities that come from geographic co-location and interconnectivity. Combined with the notion of an ongoing search for the unknown, space can take on a whole new meaning, particularly in the context of regional systems of economic and social advancement. For regional systems (be they small nation states, large cities or regional agglomerations), space is often deterministic and bounded. According to Karwandy (2008, p. 16) proximity leads to 'a non-economic structure that builds trust, common values, exchange of information and ultimately a collective development process'. The whole notion of clusters (see Chapters 3 and 4) is that they can turn short-term benefits of exchange relationships into second-order benefits that determine long-term competitiveness. This suggests that space is relative in some way. But the question is, how does space matter to innovation? Does it generate scale or scope effects? How do we measure performance in the context of space or proximity in terms of economic development and success?

The literature on regional economic agglomerations has grown to incorporate the concept of learning, especially through systems of interaction that generate tangible and useful outcomes. Lundvall (1992) argued that knowledge is the most strategic resource and learning the most fundamental activity for competitiveness. Similarly, Drucker (1993, p. 8) asserted that 'the basic economic resource ... is no longer capital or natural resources ... it is and will be knowledge'. Scandinavians and the Low Countries in continental Europe, in particular, were early adaptors and adopters of Lundvall's learning economy model of innovation. They explicitly sought to nurture interactions among players at the local level to generate shared culture and institutions. In their analysis of Nordic clusters, Coenen *et al.* (2004) and Coenen and Asheim (2004) adopted a broad definition of innovation (following Cooke *et al.*, 2003) focusing on the transformation of knowledge into novel wealth-creating technologies, products and services through processes of learning and searching.

In the context of our four Ps of innovation (prices, place, process and people), the innovation system or learning economy literature adds depth to the concept of special processes as a driver of innovation. While there is a diversity of views of what can or should be accommodated in this conceptual framing, all of the literature has some common elements. First, the innovation systems theory goes beyond the narrow, economically derived traded interdependencies of co-location to explicitly incorporate and focus on untraded interdependencies that usually are not and sometimes cannot be captured through transactions – be they 'mysteries in the air' or social capital, culture and insight, they are always tacit and frequently geographically sticky. These interdependencies are artefacts of meso-level social networks that engage individuals and organizations in collective thought and action – continuous, fluid engagement generates significant learning that can only be accessed and used by those who are part of the system. Finally, at the micro-level, institutions are no longer characterized as atomistic traders but now have personality, either in terms of their corporate cultures or personified by key individuals who are their face and voice.

THE SYSTEMATIC THEORIES OF LEARNING AND INNOVATION

Several analytical models have been developed and tested that contribute to the debate and to the literature. Prior to the 1990s, the exploration of technological performance and proximity in a policy-making context traditionally revolved around measures of inputs, such as gross expenditures on research and development (called GERD) or the number of high-quality personnel, or outputs, such as patents, publications or trainees. Although these measures have been standardized across OECD countries (OECD, 1996, 1997), it is now apparent that there were limitations to such approaches. According to the OECD (1997, p. 9), 'conventional indicators do not offer convincing explanations of trends in innovation, growth and productivity' and they ignore the value of interactions among key actors that contribute to these trends. A more convincing approach was captured through the premise of national innovation systems (NISs), which focuses on understanding these interactions and how knowledge is generated and shared and how this contributes to not only macro-productivity but to firm-level productivity (Freeman, 1987; Lundvall, 1992; Nelson, 1993; Patel and Pavitt, 1994; Metcalfe, 1995; OECD, 1997; Archibugi *et al.*, 1999).

Metcalfe (1995, p. 38) defined an NIS as comprising 'that set of distinct institutions which jointly and individually contribute to the development and diffusion of new technology and which provides the framework within which governments form and implement policies to influence the innovation process. As such it is a system of interconnected institutions to create, store and transfer the knowledge, skills and artefacts which define new technologies.' Mowery and Oxley (1995) pointed out that any definition must include more than research actors. Public programmes intended to support technology adoption and diffusion must be included, along with the array of laws and regulations that define intellectual property rights and manage the discovery, production and marketing systems.

The NIS approach acknowledges not only the key role of knowledge but also recognizes the value of a systems approach to managing and viewing technological performance, going beyond the more traditional linear approach to innovation analysis. This macro-level approach dominated in the early years, at least partly because it was developed and adapted primarily to small, homogenous, unitary states in northern Europe.

Localization of knowledge and processes became markedly important as local and regional governments, especially in larger countries and federalist states, became enamoured with the notion of local or more often regional innovation systems (RISs). Doloreux and Parto (2004, p. 2) asserted that this was preceded by the 'rediscovery [on the part of] many academics of the importance of the regional scale and the importance of specific and regional resources in stimulating the innovation capability and competitiveness of firms and regions' (further explored by Malmberg and Maskell, 2002; Asheim *et al.*, 2003; Cooke *et al.*, 2003; Wolfe, 2003; Isaksen, 2004). A natural extension of the NIS approach, the RIS framework represents a more systemic approach applied to localized gains, knowledge capture and exchange processes. This has tended to lead to the policy push for localized knowledge centres, many with a declared emphasis on specialized technological development, as a way to attract firms and high-quality personnel to regions. This localized or regional approach is designed to generate economies of scope and to leverage untraded interdependencies. Untraded interdependencies, a concept introduced by Dosi (1984) and expanded upon by Lundvall (1988, 1992), has been extensively employed by Storper (1997), who refers to cumulative-causation prone externalities, including conventions, informal rule, and habits that coordinate economic actors under conditions of uncertainty. According to Storper (1997, p. 5), these relations constitute 'region-specific assets in production' and represent 'geographical differentiation in what is done, how it is done, and in the resulting wealth levels and growth rates of regions'.

In theory, the regional innovation system and localized approaches, such as the much-referenced and globally adopted notion of clusters (Porter, 1990), are designed to reduce transaction costs for those that engage within these systems. Yet, at the heart of any of the agglomerations are the oft-ignored complexities associated with the interactions of differentiated actors, institutions, knowledge, culture and resources. These points of convergence are assumed to cultivate innovative processes and outcomes.

The natural response from a policy perspective is somehow to measure these exchanges or, at the very least, assign prescriptive or explanatory power to them. However, developing such measures or prescriptions is difficult. Several approaches attempt to capture or, at the very least, illustrate the exchanges in such systems. As with clusters, there is always some element of conceptual fuzziness or lack of consensus as to definitional boundaries and empirical approaches.

Kenneth Boulding's social triangle (presented in his 1973 book *The Economy of Love and Fear*) is one key framework that can be used to illustrate institutional interconnectivity that, theoretically, can drive the innovation process. The driving notion is that actors within systems do not operate in isolation but, rather, form bi-lateral, multi-lateral or hybrid relationships in order to optimize exchange or trade. In the case of Boulding's social triangle, interactions between the economy, society and polity can be illustrated through interlocking spheres. The framework is dynamic, acknowledging that institutional evolution and change can impact on the structures of the relationships in question.

Another similarly structured framework, the triple-helix model, emerged out of a workshop held on *Evolutionary Economics and Chaos Theory* in 1994, led by Loet Leydesdorff and Henry Etzkowitz (proceedings from this workshop were later developed into a book by the same title). The workshop was designed to explore both institutional analysis of the knowledge infrastructure and evolutionary analysis of the knowledge base of an economy (Leydesdorff and Meyer, 2006). The triple helix was originally conceived to be an interactive model consisting of 'overlapping, yet relatively independent institutional spheres' (Etzkowitz and Leydesdorff, 2011, p. 16). According to Leydesdorff and Etzkowitz (1998, p. 1441), all components of the framework are necessary in order to 'capture contemporary innovation processes'. The triple helix takes a step beyond Boulding's triangle, in that it is explicitly applied to the problem of economic innovation and attempts to introduce dynamic 'market forces, political power, institutional control, social movements, technological trajectories and regimes' wherein 'operations can be expected to be nested and interacting' (Etzkowitz and Leydesdorff, 2000, p. 113).

The triple helix model has evolved through a number of stages, closely following the economic and political regimes in place (Etzkowitz and Leydesdorff, 2000, pp. 111–112). The three evolutionary stages have been termed triple helix I, II and III. In the first configuration, the nation state was posited to oversee academia and industry and actively direct relations between them. This configuration of the model could be found in the former Soviet Union and in East European countries under the influence of communism, and remains in a few authoritarian regimes. Triple helix II, the next evolutionary step, consists of separate institutional spheres with clearly defined borders dividing them and highly circumscribed relations among each of the spheres. This is typified by Scandinavian countries such as Sweden. An increased focus on tri-lateral initiatives for knowledge-based economic development, however, has transformed the model into what is known as the triple helix III configuration, where there are overlapping institutional spheres with each taking the role of the other and giving rise to hybrid organizations at the interfaces.

A key take-away from the Ryan and Phillips (2004) study that used the Picciotto (1997) model and Boulding social triangle is that advanced-technology-based communities or regions are greatly influenced by the efficacy and efficiency of the embedded networks and interactions of actors (from various sectors) within a cluster. The stronger and more effective the relationships or connections within the system are in terms of accessing and sharing knowledge, the more competitive that region will be in the context of a broader global market. Ouimet *et al.* (2007), in a complementary analysis of the photonic sector in Quebec City, concluded that firms that are more engaged and connected in the innovation system are more likely to realize drastic innovations. The general analytical framework takes into account the notion of both formal and informal interactions, as well as factors such as resources, knowledge, high-quality personnel, impact on regional innovation and productivity. This includes both local and global factors. Managing knowledge in a cluster setting requires not only proficiently handling the patent portfolio (know-what knowledge) but also includes the proactive process of 'amassing, integrating and leveraging tacit knowledge' (Ryan and Phillips, 2004, p. 230). This means sourcing not only brain power (i.e. high-quality personnel) but also leveraging key non-codified knowledge (e.g. culture and practices) that is embedded in institutions. At the intersection of regional actors and institutions, where respective activities converge, idea generation is presumed to occur.

Phillips (2001, 2002) offers a more detailed framework for exploring the dimensions of what the author views as an *entrepôt* configuration. The notion of *entrepôt* trade has largely been expunged from current economic literature and, according to Phillips (2001), can only be found in more dated pieces – 25 years or more – where trade studies focused exclusively on goods. *Entrepôt*, at the time, was defined as 'a centre at which goods are received for subsequent distribution ... [which] has facilities for the transhipment of imported goods or their storage prior to their re-export, without the need to pass through customs control' (Bannock *et al.*, 1972). According to Phillips (2002, p. 41), the *entrepôt* can best be examined by looking at 'its relative role in creating knowledge, using knowledge and commercialising new products'. Phillips (2002, p. 31) asserted that the Saskatoon-based, oilseeds cluster he studied exhibited many attributes of a classical *entrepôt*, with 'the majority of the basic research and many of the proprietary technologies being imported, assembled into new crop varieties for commercial release first in Western Canada, and then exported on a largely tax-free basis as intermediate product to global markets'. Canada, more broadly speaking, can be viewed as the main canola innovator based on its history as an early developer and adopter of all the new traits over the past four decades. However, it is important to note that, in the context of the *entrepôt* approach, much of the applied research to develop the processes to create new canola varieties was done in other countries (such as the USA, the UK, Belgium and Germany), while much of the applications-based or value-add research was happening in other parts of the world. The Saskatoon-based *entrepôt* (outlined in further detail in Chapter 6) undertakes and assembles the know-why, know-how and know-who of varietal breeding and primary production while the lion's share of the activities upstream and downstream in the production system are done elsewhere.

Phillips' *entrepôt* model quite neatly takes into account the notion of Bathelt *et al.* (2004) that 'local links and global pipelines' drive regional and national systems of innovation. The pipeline–buzz framing (or the notion of local/global as explored further by Gertler and Levitte, 2005) acknowledges that, while place and proximity matter in terms of innovation, the activities of a given region or cluster cannot be cut off from the rest of the world. In fact, the capacity for regional innovation often depends on the leveraging and extraction of global sources of knowledge and resources, such as the case of the Saskatoon canola-based innovation cluster.

Drilling down into the components that underlay these frameworks, we find that knowledge and knowledge-based activities can be moving targets. Evolution of knowledge requires learning; learning requires engagement; engagement involves people who often may be mobile in continually evolving firms and institutions. Needless to say, this is a nebulous space. In practice, there is rarely a question of whether one works with exclusively tacit or exclusively codified knowledge. It is the overlaps and complementarities between both types of knowledge that are most important (Nonaka and Takeuchi, 1995).

The learning economy itself is largely a concept of the RIS approach, which appeared in the early 1990s (Cooke, 1992, 1998, 2001), growing out of Lundvall's (1992) concept of the NIS. This approach emphasizes the importance of interactive learning and the role of institutions in explaining innovation performance and growth across various countries. Regions are viewed as important mediators for governance and coordination between the national and local levels. Lundvall and Borrás (1998, p. 39) suggested that 'the region is ... the level at which innovation is produced through ... networks of innovators, local clusters and the cross-fertilising effects of research institutions'. While the learning approach represents a softer and broader approach to innovation than traditional economic or strategic management approaches, it frequently suggests untapped growth potential. Lundvall (2004) argued that the model necessitates institutional reforms and organizational change to promote learning processes, while Drucker (1999) suggested that it might demand a new more radical economic policy.

Lundvall *et al.* (2004) recognized the tension between two particular modes of learning for innovating firms. First, science-based learning is based on the production and use of codified, scientific and technical knowledge, referred to as science, technology and innovation (STI) learning. Experience-based learning, or learning by doing, using and interacting (DUI) is comprised of informal, non-codified exchanges in efforts to mobilize tacit knowledge in order to problem solve (Jensen *et al.*, 2007). When combined effectively, these two modes of learning are deemed to be highly complementary and can amplify firm-level innovation. The DUI mode helps to alleviate bottlenecks and increase efficiencies in a given organization as it attempts to contribute to and draw from the broader STI system.

At this point, it is important to explore in a bit more depth distinctions in the literature around innovation systems. Coenen *et al.* (2006) explore these distinctions in comparative case studies of the Saskatoon agro-biotechnology cluster with the Skanian region in Sweden, focusing on knowledge archetypes and artefacts that are core to the respective regions (Asheim and Coenen, 2005, also explored 'knowledge bases' in Nordic clusters). In essence, the study provided a socio-spatial analysis between regions that appeared to have some knowledge-based commonalities. The innovation systems literature supports the notion that innovation and innovative processes are carried out through a network of assorted actors, often comprised of industry, universities and (multilevel) government (Etzkowitz and Leydesdorff, 2000). This structure of innovative activity is underpinned by an institutional framework (Edquist, 1997). The empirical studies of national NIS and RIS explore the common habits, routines, established practices, rules and laws that regulate and provide incentives for interaction among this heterogeneous group of actors (Edquist and Johnson, 1997).

Key to both NIS and RIS is the continual emphasis on inter-connectivity: interaction not only between actors but between institutional paradigms. This provides scope for complementarities among these various actors and 'factors'. Innovation systems are distinguished in a variety of ways and such distinctions rest heavily on the nature of boundaries – geographical boundaries (i.e. local, regional or national) but also sectoral and technological boundaries. The cluster paradigm brings something different to the mix – a configuration of firms or organizations in the same or adjacent industrial sector(s) that are concentrated in a small geographical area. The location is the common ground between regional innovation systems and clusters. That said, overlaps should not be overemphasized. By definition, a regional innovation system can stretch across many sectors – and clusters – in the regional economy, provided that the firms and knowledge organizations interact in systemic manners.

Neoclassical economics traditionally reduces knowledge to information that is ubiquitous and free flowing. The RIS and cluster literature conceptualizes learning and knowledge exchanges between various actors in an economy as inherently social processes that embody value (Lundvall, 1992). Furthermore, the RIS approach draws substantially on the notion of complementarities that grow out of recurring interactions and those 'untraded interdependencies' that are often linked to the region in the form of conventions, informal rules and habits (Storper, 1997). The primary rationale behind the regional perspective is that it facilitates bottom-up analysis of the interactions and inter-organizational learning processes (Doloreux and Parto, 2005). The role of embeddedness is important here. Personal and professional relations and networks that can trigger economic action and outcomes are embedded or ingrained in the social and cultural context of a region or cluster (Granovetter, 1985). In contrast to the NIS, which has a bit of an autarkic flavour, the RIS approach acknowledges the need to combine both local and non-local knowledge (through pipelines) as well as to mobilize local skills and competencies (the buzz) to realize benefits beyond that of the region alone (Doloreux, 2004). The local buzz–global pipeline metaphor underscores the significance of non-local connections for exploiting new and valuable knowledge created in other parts of the world. The facility to link local and non-local capacity helps clusters or regions avoid cognitive and economic lock-in.

The industrial knowledge base provides an important dimension to deconstruct and compare regional innovation systems (Coenen *et al.*, 2006). Brink *et al.* (2004) asserted that the knowledge base includes a specific area of knowledge as well as its embodiment in techniques and organizations. Scott (1995) concurred, offering a threefold categorization of these processes of knowledge management as regulative, normative and cognitive rules. Studies of RISs seem to be mainly geared towards analyses of regulative (e.g. formal rules, laws, standards and governance structures) and normative (e.g. values, norms and codes of conduct) rules. In comparison, there have been very few studies of what some call sectoral innovation systems, which pay more attention to cognitive institutions such as problem agendas, paradigms, categories, classifications and search heuristics. Nelson and Winter (1977) argued, for example, that engineering practices in firms are less sensitive to price-related changes on the market than to technical ideas and beliefs about where to go, what problems to solve and what sort of knowledge to draw on.

The evolving theory of innovation systems focuses on the systemic effects of institutions and individuals applying science and learning by doing. The focus tends to be on the diversity and array of connections between institutions and individuals and on the conceptual theories of learning. All of this then is juxtaposed into specific geographical situations. The innovation systems literature offers significant insight into the role of special processes but suffers from a largely under-theorized role for firms or individuals. Just as the cluster theories are weakened by a poorly described role for firms, the innovation systems literature says little about how firms benefit from these non-market or extra-market social processes that create untraded interdependencies.

THE RESOURCE-BASED VIEW OF THE FIRM AND INNOVATION SYSTEMS

Lost in the shuffle of innovation and regions is the important role of the firm in innovating. While the DUI-STI modes of learning and exchange offer tantalizing glimpses of how firms might access and utilize resources to achieve innovative outcomes, the formal link is seldom explored.

Innovation systems can be linked to business literature through the externalities perspective (Marshall, 1890; Lucas, 1988; Romer, 1990, 1994; Krugman, 1991b; Krugman and Venables, 1996), which posits that proximity fosters indirect linkages or an informal network that enables location factors and public goods to enhance the local division of labour (specialization), increases the presence of spill-overs and improves the overall technology development process. Externalities, also known as location factors, are the aspects of a place that exist beyond the realm of any single firm and the transactions they engage in yet remain spatially contained so that access is restricted to individuals or organizations within an appropriate proximity to that location. They range from being readily observable elements such as the labour pool or service infrastructure to more intangible items such as education systems or the quality of local leadership. Maine *et al.* (2008) suggest that proximity to a cluster is associated with growth of new technology-based firms. According to the authors, this matters most when these firms rely heavily on broad, downstream supply chain effects in order to enhance their resources and capabilities. That said, firms appear to differentially benefit from co-location and proximity to a cluster.

The foundation of this perspective is that, even though location factors or externalities exist everywhere, they deliver the most value and form the basis of competitive advantage in locations where organizations foster the *right* relationship structure. Advantages derived from location factors are much harder to imitate or recreate because they are rooted in the involvement of multiple organizations (Barney, 1995). In this perspective, firms are linked indirectly to one another through their mutual interest in deriving value from locational factors. As a result, proximity underpins the relationship between firms and location and not just the firm-to-firm transactions emphasized in the competitiveness perspective discussed in the context of clusters in Chapters 3 and 4. Furthermore, there is no assumption that firms must cooperate, at least beyond what makes sense to them in a self-serving, competitive environment. Nevertheless, firms are inclined to protect their investment in the advantages of a particular location by using the locational advantages in a way that keeps from depleting them.

A core difference between the externalities perspective and the competitiveness approach discussed in Chapter 3 is the way in which a firm is believed to respond to the environment. The competitiveness perspective assumes that uncertainties from the external environment can be managed by internalizing them through the development of formal relationships and long-term contracts – in short, through the market. The externalities perspective, in contrast, emphasizes that a firm designs and builds its internal capabilities with the intent of proactively responding to the external environment. Hence, location factors are valued for the ways in which they enable firms to interface with the broader external environment. For example, a thick labour force allows the timely redevelopment or deployment of skill sets, while spill-overs support early awareness of changes and the dispersion of sagacity. When the basket of location factors is just right, there will be a convergence of interested firms; in other words, demand for a location will grow and agglomeration can emerge.

The externalities literature is usually traced back as far as Alfred Marshall's (1890, 1923) work on localization economies. The key aspect of Marshall's work that pervades the externalities perspective is the notion of 'mysteries in the air', the phrase he used to characterize the pure external effects of specialized knowledge and processes that emerge around the core line of business in an area. The basic argument is that, over time, a unique local ability to understand and develop new ideas associated with local areas of focus can emerge. This can manifest itself in a wide variety of ways (e.g. in the quality of skills development programmes or educational systems) but overall represents a unique business sense or sagacity that can impact on the path of future development. This concept is probably best illustrated by the idea that a tradesman's (e.g. fisherman or carpenter) child who grows up exposed to the trade will have a more intuitive and deeper understanding of that trade than someone who has simply studied the skills. This simple idea of immersion or learning by doing underpins the DUI concept of learning and is widely considered a critical part of the function of the STI space.

As the externalities perspective has continued to evolve, 'mysteries in the air' have become known more as knowledge spill-overs or dynamic information externalities. According to the new economic geography literature (Krugman, 1991a,b, 1998a,b), as well as the endogenous growth literature (Lucas, 1988; Romer, 1990, 1994), knowledge spill-overs and other forms of information externalities are important determinants of long-term economic growth. Location factors related to knowledge and information appear to be more important to competitive advantage and overall performance than other location factors, such as the thickness of the labour force, access to infrastructure, and extensive forward and backward linkages. These other location factors may facilitate the emergence and extension of the cluster, but the 'mysteries in the air' are assumed to fundamentally influence firm performance.

Even though the importance of knowledge spill-overs seems to be pervasive, the literature is full of debate and theories about when, where, how and why dynamic externalities work. In Romer's endogenous growth model (1990), knowledge flows are hypothesized to have a negative impact on long-term growth – the dissemination of ideas without offsetting compensation through imitation, spying and mobility of labour is assumed to discourage investment because firms cannot be certain that they will be able to appropriate enough value to recoup their investment before imitators enter the marketplace (Bekele and Jackson, 2006, p. 16). Others assert that firms embedded in regional and national innovation systems gain more than they lose through these processes (e.g. Lundvall, 1992). Beyond these basic orientations, there are a number of unresolved debates around whether it is better for firms and regions to diversify or specialize their knowledge advantages, whether monopoly or competition in knowledge generation is best, what level of knowledge sharing is most effective (e.g. local, regional, national or international) and the relative importance of knowledge in traditional sectors versus so-called high-technology sectors.

These externalities are then accommodated in firms through each individual firm's strategic policies and processes. Penrose's (1959) resource-based view (RBV) of the firm, which focuses on the internal elements of a firm (in contrast to the externally focused PBV discussed in Chapter 3), posits that firm managers will first look inside to consider whether a firm has what it takes to be number one. This perspective recognizes that firms evolve over time and that there are points in a firm's life cycle that are better suited to pursuing dominant market positions. In contrast, the PBV identifies dominance as the only desirable competitive position for a firm. The contrast between these views establishes a chicken-or-egg argument;

does an external focus (PBV) generate the ideal internal structure or does a meticulous focus on the internal structure (RBV) generate desirable external accomplishments? The answer to this question is less important than the connection it underscores between a firm's internal and external environment.

The underlying principle of the RBV is that the firm is an evolving collection of resources (Penrose, 1959), suggesting that, over time, firms will change because of their ability to learn or acquire new resources and capabilities. The strategic goal, from the perspective of the RBV, is to secure the factors required to build core competencies. Core competencies become the basis for establishing and maintaining competitive advantage. Strategically, those firms that do not develop their resources and capabilities fall behind and ultimately fail. Thus, the long-run focus of the RBV means that innovation is pursued proactively in an effort to disrupt the market in favour of the innovative firm (as posited by Schumpeter, 1939). In contrast, firms competing from the PBV consider innovation as an incremental process, necessary only to react to the changing needs of the market – in short, consistent with the induced theory of innovation discussed in Chapter 2. This is not to say that the RBV leads to a supply–push model but rather that it seeks to anticipate the changing needs of the market rather than simply react to them.

The RBV of the firm argues that changing demand factors and exogenous economic shocks are not inherently good or bad. Threats and opportunities are a function of a firm's internal capability. Core competence or incompetence determines whether or not an external event becomes a threat or an opportunity for a firm. For example, the design and marketing capabilities of Japanese car makers allowed them to do a better job of responding to the OPEC oil shocks in the 1970s, thus turning what was a threat to North American car makers into an opportunity for themselves. For this reason, the RBV argues that competitive advantage is derived from a firm's internal attributes. Barney (1995, p. 50) identified four categories of resources and capabilities: (i) financial; (ii) human; (iii) organizational; and (iv) physical. Firms theoretically possess assets and liabilities in each of these categories. The RBV of strategy suggests that the firm has direct control over the possession, maintenance and development of all of these assets and liabilities and must take the initiative to utilize each.

Despite being valuable, resources may fail to provide a competitive advantage if they are not rare. If most competitors can access a similar valuable resource, no one will retain an advantage. However, even when valuable resources are common, they remain important to competition. Without the valuable resource, a firm would be at a disadvantage. When firms possess a valuable resource that is also rare, they must find ways to protect their advantage. Hence, the characteristic of imitability becomes important for achieving a sustained competitive advantage.

Firms must find ways to keep competitors from duplicating the valuable, rare resource or substituting a strategically equivalent resource. Barney (1995) identified three ways in which resources become more difficult to imitate. When resources develop under unique historical conditions, they can come to reflect the unique personality, experience and relationships of the firm. Alternatively, resources or capabilities may incorporate an intricate web of many small decisions. The individual decision may not be difficult to imitate, but the fact that they are so minor makes them invisible to other firms and thus impossible to imitate. Finally, resources based on social complexity are more difficult to replicate than standard physical resources commonly protected by patents. Reverse engineering can quickly solve the mysteries of physical resources but is ineffective against 'organizational phenomena like reputation, trust, friendship, teamwork and culture' (Barney, 1995, p. 55).

At the core of the RBV is the assumption that developing, controlling and ultimately combining resources and capabilities into novel products, processes and services can provide a greater return than constantly jockeying in pursuit of incremental increases in efficiency or market share as in the PBV. In this way, firms pursuing the RBV are much more interested in assuming the risk and cost of creating markets than firms adopting PBV. Rugman and Verbeke (2002, p. 771) observed that 'the resource-based view of the firm implies that firms pursue disequilibrium (monopolistic advantages) through a process of Schumpeterian competition, path dependencies, first-mover advantages, irreversible commitments and [use of] complementary or co-specialized [resources]'. Firms with this focus on innovation and market making require access to a diverse set of resources and the capability to engineer or interface these building blocks to create new products, processes and services. These firms also need ways to ensure that their novel creations remain valuable, rare and inimitable. Finally, the operation has to be optimally organized to maximize the market returns of innovation. From the resource-based perspective, economic agglomerations are strategically relevant for the ways in which they can support the process of innovation and market making. This theoretical framing of the firm should, but has so far failed to, provide insights and evidence of the relative importance of different innovation system factors on the efficient and effective transformation of inputs into innovative outputs and ultimately economic outcomes.

METHODS

Given the rather abstract nature of relationships and learning operating in a regional economy, it can be extraordinarily difficult to map or empirically test innovation and the efficacy of interactions and exchanges that underlie innovative activity. Jensen *et al.* (2007, p. 684) pointed out problems in empirical approaches, suggesting that quantitative survey-based studies that attempt to support and justify policy demonstrating bias towards elements of 'purely formal scientific and technical knowledge and formal processes of R&D'. This leads us to the examination of standards that have driven these empirical approaches or enquiries. In particular, the Frascati approach, which targets ways to measure R&D, and the Oslo manual, which focuses on innovation measures, are often applied in studies of clusters and regional innovation.

Developed by the OECD and the National Experts on Science and Technology Indicators (NESTI) group in 1963 in Frascati, Italy, the Frascati manual deals with measuring the resources devoted to R&D in high-performing sectors. The definitions provided in this manual have become internationally accepted and played a key role in developing an understanding of the relationship between science and technology and economic development. Until the 1993 version of the manual, there really was no precise way to sample a population of firms, no matter what their size. Member countries, as a result, would interpret the guidelines differently, which put a wrench into international comparisons. Two approaches were promoted by the 1993 version of the manual, wherein researchers could choose to either extract a sample set from an entire population of a given sector or they could survey only those firms that supported R&D (OECD, 1993, p. 107). Over the past 40 years and five editions (the last published in 2002), the OECD and the NESTI group have developed a whole range of documents on innovation known as the *Frascati Family*.

The Oslo manual on innovation is considered to be a part of this family (OECD, 2005). The first edition of the manual, issued in 1992, showed that it was possible to map the complex and differentiated processes of innovation. The third edition of the manual outlines factors of innovation at the firm level only, covering four types of innovation: (i) product; (ii) process; (iii) organizational; and (iv) marketing. Furthermore, it provides guidelines for collecting data on the general process of innovation (e.g. activities, expenditures and linkages), the tracking of significant changes in the firm (e.g. processes), the factors that influence innovative activities (e.g. incentives) and the outcomes of innovation (e.g. performance). The manual has gone a long way to normalize the concept of innovation along the lines of Schumpeter (1939), asserting that innovation is 'the implementation of a new or significantly improved product (good or service), or process, a new marketing method, or a new organizational method in business practices, work-place organization or external relations'. The guidelines provided in the Oslo manual can be used as a way of informing government as to how policy can be used as a lever to foster innovation. However, the OECD recognizes that innovation is complex and that there is no definitive model of innovation and, hence, no single prescribed course for government to take. Better indicators need to be developed that can measure the innovative performance of firms.

The Frascati and Oslo manuals have been translated into a range of surveys and databases. Gault (2010) suggested that, at root, all of the innovation metrics focus on inputs, activities, linkages and outcomes. The basic input measures – GERD, business expenditure on R&D (BERD) and the sources of funds and performers of effort – combine with a set of quantitative and qualitative questions about the activities – Who are the actors in the system? What is the nature of the activity engaged in by the actors, including the cost to the actor? What knowledge is produced? What knowledge is used? Where is the activity happening – in a region, sector or institution? Why is the actor doing the activity? – to define the scope of measurement of innovation. In addition, there is an effort to measure the ways that expenditure, materials, energy, human resources, capital investment and knowledge are transmitted through social organizations, supporting infrastructures and discipline networks. Finally, metrics attempt to address the outputs and outcomes, including the change in market share, skill levels of the workforce, patents, publications, new products or processes. These questions can be directed towards specific communities (i.e. community innovation surveys or CISs), technologies (i.e. technology use surveys that test for specific applications, such as ICT) and firms or sectors (i.e. management practices use surveys).

A range of official surveys have been or are regularly undertaken by statistical and government agencies: the CIS in Europe covers all 27 member countries and there have been innovation surveys in most other OECD member states, in China, Russia, Africa and Latin America and, in 2009, one began in the USA. A number of scholarly groups have also undertaken to apply the survey methodology to specific innovation systems. There is an important data dimension to the Innovation Systems Research Network (ISRN), which in 2000–2011 examined the impact and importance of innovation in 27 centres across five Canadian regions both in newly emerging knowledge-intensive areas and in more traditional sectors. Data for the project was gathered through a survey process administered across the country. More than 1500 surveys were administered to public and private sector actors across the 27 hypothesized clusters or innovation systems over the decade. More than 400 of these surveys focused on the life science and biotechnology area. The data consisted of a blend of both quantitative and qualitative elements, generally collected over a survey period of up to 2 years in each regional system. The survey instrument covered not only factual information associated with firm size and history, but also explored strategies, networking and relationships, locational benefits and infrastructure, and the role of the cluster in firm performance. While there was hope that these surveys might provide a statistically robust set of metrics, the distributed nature of the surveying and the diffusion of the data made it difficult to do comparative work among the 27 data sets – most of the local and regional data was used to amplify and ground narrative analyses of the discrete innovation systems and not to test the theoretically grounded hypotheses.

In spite of significant efforts to assemble an appropriate set of metrics, problems remain. A few examples of recent efforts to fill the gaps help to illustrate the difficulties. The Institute for Competitiveness and Prosperity (ICAP), a Canadian-based organization tailored after the Porter-led Harvard Institute for Strategy and Competitiveness, has a searchable database of centres that may exhibit attributes of innovation systems. The database includes entries on employment, wages and location quotients for 41 identified industry clusters producing traded goods, as well as data on local and natural resource industries for each of Canada's 25 city regions (i.e. census metropolitan areas or CMAs), the ten provinces and the nation as a whole. Unfortunately, this data is delineated only by industry and by province or CMA. This limits the ability to focus on the system level because their boundaries are inherently fluid and are not necessarily bounded by standard political or jurisdictional boundaries. Most sources of allegedly cluster-based data in Canada and other countries suffer from this defect. Other than CMAs, there are few or no subprovincial data available. Additionally, data by sector are limited to that which is delineated through North American Industry Classification System (NAICS) codes. In knowledge-based industries such as biotechnology or ICT, data collected through NAICS codes is often incomplete and highly subjective. Furthermore, there is no consensus on where to draw lines or pull together multiple codes to best represent these mode 2 type systems. Boundaries, particularly in advanced technology areas, cannot be defined by conventional product-based, industrial or sectoral boundaries.

One response to the problems associated with using NAICS and CMA data was attempted within the ISRN project. A concordance was formulated between the Statistics Canada Survey of Innovation (a collection of surveys carried out and reported on in 1993, 1996, 1999, 2003 and 2005; http://www.statcan.gc.ca/pub/88f0006x/2009002/aftertoc-aprestdm2-eng.htm), the Biotechnology Use and Development Surveys (http://www23.statcan.gc.ca:81/imdb/p2SV.pl?Function=getInstanceList&SurvId=4226&SurvVer=2&InstaId=14687&SDDS=4226&lang=en&db=imdb&adm=8&dis=2) and the ISRN database. The objective was to link the NAICS/CMA data to geographical clusters to illuminate their local and global relationships and to determine firm innovativeness and performance as identified in the ISRN company and institutional surveys. The Survey of Innovation, part of an ongoing programme to measure innovation in Canada, is conducted on approximately 400 firms every 3–4 years, covering a 3-year reference period. The Biotechnology Use and Development Survey provides further statistics on biotechnology, targeting companies developing new products and processes using biotechnologies.

In spite of heroic efforts to generate data, most innovation system assessment methodologies in Canada and elsewhere appear to lack elements that would be necessary for a compelling analysis. Many approaches tend to be narrative in nature and focus merely on the institutional form of the system to the neglect of emphasizing more causal elements associated with system development and success.

The OECD (2010) recognized the limits of current data and engaged a blue-sky group to assess the gap in the data. This group concluded that research on innovation is currently fragmented and that indicator and econometric research needs to move from inputs and activities to outputs and outcomes, which will require new and better data that assess how the actors fit together in systems. The OECD asserts that design of innovation policy must take into account the characteristics of technologies, people and locations, as well as the linkages and flows among them. To this end, new methods of analysis are needed to understand innovative behaviour, its determinants and its impacts at the level of the individual, the firm and the organization. In pursuit of that goal, they identified and aggregated the known evidence on 50 traditional and non-traditional measures of innovation. A quick perusal leaves the distinct impression that, while all of the measures may have value, there is little convergence among OECD member states on when and what to measure, with the result that virtually none of the non-traditional measures are comparable, which seriously limits their value for comparative policy analysis.

CONCLUSIONS

While the innovation systems theory, at both the national and the regional level, is more subtle and adaptable than place-based cluster theory, it suffers from many of the same challenges. The incorporation of institutions, incentives, relationships, knowledge and learning by doing as hypothesized causal variables clearly adds more flesh to the bones of innovation theory but also risks making the theory over-determined. In its current form, it is difficult to see how one could set up experiments with refutable hypotheses that would allow us to test scientifically the value of the different aspects of the theory. Moreover, the almost complete absence of any discussion of the role of firms (and firm strategy) or the role of individuals (except perhaps as relatively anonymous nodes in a social system) seriously limits our ability to follow the causal pathways.

The complexity in the theory undoubtedly complicates data collection and analysis. The result has been a many case studies and much qualitative analysis but limited comparative or econometric analysis.

6 Policies, Strategies and Metrics Related to Innovation Processes in Saskatoon

Malmberg and Maskell (2002) note that when empirical research showed that firms in a localized system often did not conduct much business together, the study of business-to-business linkages gave way to the study of knowledge spill-overs. Bekele and Jackson (2006) identified a number of studies that highlight the presence and value of social networks, inter-personal relations, face-to-face encounters, casual or tacit information flows, and norms of trust and reciprocity among local actors. Often these interactions are tracked through artefacts, such as patents, publications and citations. Storper and Venables (2004) suggested that face-to-face contact is the causal mechanism that determines the effectiveness of interaction and thus the degree of knowledge spill-over or transfer. They argued that face-to-face communication is not just an exchange; rather, it is a performance, where speech, actions and context combine to lead to complex communication on multiple levels at the same time. This work implies that there is a varying quality to interaction. Simply bringing companies, their employees, local scientists, consultants and other support agents into contact does not guarantee the benefits of agglomeration.

This chapter examines interactions and spill-overs in the context of the Saskatoon economy, using a range of analytical approaches. Overall, the evidence suggests that STI exchanges, involving codified knowledge, may be the most readily observable but are perhaps less important for the effective functioning of the innovation system than the DUI subroutines.

MAPPING INNOVATION PROCESSES

In contrast to Porter's competitiveness diamond, which illuminates institutional relationships embedded in special places, the innovation systems approach, which focuses on special processes, requires a new organizing heuristic. While some have used the triple helix, it tends to be overly focused on institutions. Bathelt *et al.* (2004) suggested that instead we should focus on global pipelines–local buzz, which in Saskatoon was translated into an operational approach through the concept of an innovation *entrepôt*. This approach tracks both codified and tacit knowledge flows through their artefacts, including the usual publications, patents and citations but supplemented with evidence related to personal service contracts, formal and informal business interactions and regulatory actions. Burt (2005) asserted that informal networks are a functional component of business strategy and are the key to amassing social capital.

Phillips (2002) examined the agricultural biotechnology community in Saskatoon through the innovation *entrepôt* concept, focusing on the processes and systems involved in creating knowledge, using knowledge and commercializing new products resulting from those knowledge systems. On the face of it, the basic artefact of the number of new innovations and new canola varieties produced by Canada (led by teams and firms in Saskatoon) suggests that Canada is the main global innovator in this product category. However, as already discussed, a deeper analysis of the knowledge flows and processes revealed that a significant share of the applied research to develop the processes to develop these varieties was done in other countries and much of the applications-based research (e.g. uses for new oils) happened elsewhere. This suggests that Canada instead operated in a niche in this global knowledge-based industry – as an *entrepôt* undertaking and assembling the know-why, know-how and know-who of varietal breeding and primary production (Malecki's typology of knowledge was discussed in Chapter 1) – and that the bulk of the activities upstream and downstream of this stage in the production system were and may continue to be done elsewhere. Figure 6.1 illustrates the relationships between the global industry and the Saskatoon *entrepôt* where the innovation system is examined in the context of knowledge creation and knowledge commercialization, which involves related and supporting industries, regulators, producers and consumers.

In the context of knowledge generation, at the know-why level, Canada is the single largest country producing pure agricultural research into canola, but the basic research this underlies that effort is disproportionately coming from the USA (35%) and other countries. Canada is the only major canola-researching country that imports significant amounts of basic research relative to its canola research (Phillips and Khachatourians, 2001). Moving forward in the research stream, Canada, and particularly

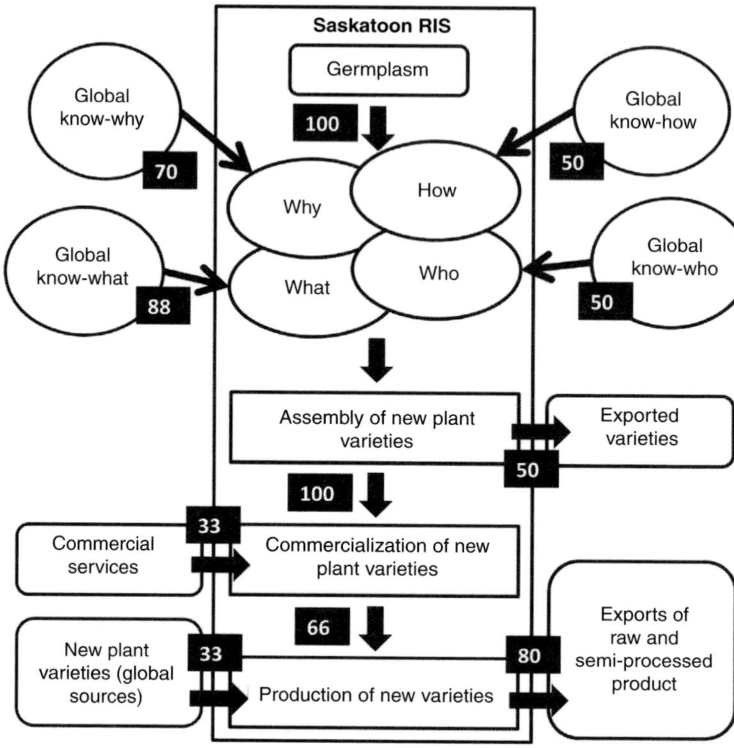

Fig. 6.1. The Saskatoon biotechnology *entrepôt.* Numbers demonstrate the relative percentages of activity across all factors, illustrating the relationship between local and global resources.

the public laboratories in Saskatoon, held a dominant position in canola-specific research until the early 1980s, with all Canadian locations combined contributing between 33 and 40% of the pure agricultural research on canola in the world up to that time. In the 1990s, the global research volume expanded greatly while Canadian research efforts were relatively steady. The result was that canola research disseminated into a wide variety of locations and institutions and that Canada's share of this research dropped to only about 20%. Furthermore, as the basic scientific research became more specialized, collaborations played a greater part in the global evolution of research capacity. The top nine canola research countries undertook between 5 and 22% of their research through international collaborations. Canadian researchers undertook only 8% of their effort via international collaborations (the lowest of all countries except India), suggesting that Canada was more self-sufficient than many of the rest at this stage.

Despite this, it is evident that Canada was far from self-sufficient. Canada has been significantly more dependent on foreign actors for proprietary know-what technologies used in the industry than for basic R&D. The Canadian patent database shows that between 1985 and 1999, there were 634 patents issued for canola-related inventions (45% for processes), the vast majority of them to foreign, mostly US-based, research teams and companies (Phillips and Khachatourians, 2001). Canadian researchers accounted for only 75 patents, more than half of them to a few federal research scientists in the AAFC and NRC. In practical terms, virtually all of the key transformation technologies used by the canola sector were the proprietary products of non-resident companies (Phillips, 2000).

Moving downstream in the product development system, field trial data provides insights into who was undertaking research, where and with whom during this critical period. International field trial data showed that Canada had an early and dominant lead over any competing country. Although 12 countries had field tested at least one transgenic variety between 1988 and 1995, Canada accounted for 54% of all field trials undertaken and did almost four times the number of trials of France, the next busiest country.

One might be inclined to conclude, based on the above data, that the regional system of innovation centred on Saskatoon is a chimera. On the contrary, the generation and transmission of non-codified knowledge in the regional system would appear to be the key factor holding things together. People develop skills and working relationships, which together convert bits of information into operable knowledge. This tacit type of knowledge is learned almost exclusively through experience. Jensen *et al.* (2007) posited in the theory of DUI that researchers learn how to do things and who to work with through trial and error. Most of the innovation literature assumes that this know-how and, perhaps more importantly, this know-who evolves within corporations or institutions. That may hold true in an industry or within firms that are largely self-sufficient, but, as noted, there are few firms that have the internal capacity to undertake all the research and development necessary to create a marketable crop variety. Some companies may have that capacity within their global operations, but in many cases working through the geographically dispersed multiple layers of these MNEs is more complex and less cost-effective than buying in from a more accessible and timely local source. Hence, although Monsanto and AgrEvo, for instance, both had giant research 'universities' and laboratories at their headquarters in St Louis and Frankfurt, respectively, both collaborated extensively in Saskatoon with both the AAFC and NRC. Furthermore, in knowledge-based industries, training and upgrading are critical, making it essential for private researchers to interact with the broader research community. For all these reasons, most of the firms in the industry developed an extensive community of networks with both collaborators and competitors, involving other private companies, the University of Saskatchewan, AAFC and the NRC.

A survey of canola firms in Canada and globally undertaken in early 1998 highlights the importance of the proximity of competitors and collaborators as factors in locating their research efforts. Half of all the respondents, representing the majority of private companies responding, acknowledged the importance of proximity to either collaborators or competitors. About 40% recognized the importance of being close to their collaborators, particularly the NRC and AAFC in Canada and key research universities in the US (Table 6.1).

As with most communities, proximity matters. Formal and informal face-to-face meetings and working side-by-side on laboratory benches and in the greenhouses are critical elements of both developing know-who and transmitting know-how. It is highly unlikely that the community would have developed if there were only competitive firms in Saskatoon; the non-competitive environment offered by the AAFC and NRC created a platform for these relationships. Both AAFC and NRC had extensive arrangements with each other, public universities and private companies. In 1995–1996, at the height of the canola research effort in Saskatoon, the NRC had more than 31 arrangements – ranging from research agreements to collaborative work agreements and licences – that brought more than 65 guest researchers from other institutions into the NRC laboratories (National Research Council Canada, 1997). In 1997–1998, when new products were just beginning to take hold in the market, the NRC welcomed 109 guest researchers. The key feature of these arrangements was that the core research team at NRC was able to learn from all of the collaborations, thereby adding further to know-how and providing a visible, efficient point of entry to access know-who knowledge. Even firms not resident in Saskatoon developed extensive links to gain access to the knowledge in those two institutions, which suggests that spill-over benefits from the know-how and know-who located in Saskatoon may have been both significant and very mobile.

Table 6.1. Key factors in decisions to locate in Saskatoon or elsewhere ($n = 28$) in a survey of canola firms in Canada and globally. (From Phillips, 2001.)

	%
Proximity to competitors or collaborators	50
Proximity to collaborators	39
Proximity to competitors	29
Access to local pool of skilled labour	25
Access to large and accepting farm market for seeds being produced	21
Key scientists either in your company or in partner organizations	18
Role of government agencies (federal, provincial, regional) related to hospitality, red tape (or lack of)	18
Access to laboratories, greenhouses and test fields	14

A key element in pulling the research effort together is specialized, skilled labour. The gradual agglomeration of canola research in Saskatoon and Canada suggests that there should be some evidence of a speciality labour market evolving. More than one-quarter of companies responding to the canola industry survey, and the clear majority of private companies, indicated that access to a deep, local, skilled labour pool was important. As noted in Chapter 4, the post-secondary educational institutions have specialized and graduated most of the lower-skilled people and a plurality of the senior scientists in the industry. While conducting research on the social dynamics of innovation at the city-region level, we asked employers where they sought and found new employees and discovered that about one-third of companies got new employees from non-local markets, with firms relying more heavily on external recruitment for marketing, management, and science, technology and engineering research staff. Local markets were more important for production and design workers. The prevailing view of many is that, while Saskatoon has a dynamic economy and attractive local labour market, there are gaps in the market that limit any individual's ability to ascend the corporate ladder locally – many feel that they need to spend part of their career away to gain skills and experience not available in the local market. This is reflected in the nearly 40% of University of Saskatchewan graduates who have migrated outside the province (Phillips, 2000). Ultimately, mobility of labour is a vital component of the innovation system – the higher the degree, the greater the mobility and intra-country and cross-national movement of employees.

One could conclude from this analysis that, during the knowledge-creation phase, Saskatoon and Canada were significantly dependent on global markets for know-why and know-what but that they were increasingly self-sufficient as research became more applied.

Theory suggests that the special processes of innovation may be as or more important in the commercialization phase of new technology and product development. Economic theory indicates that firms locating where there are extensive backward linkages into supporting industries and forward linkages into the market can realize scope economies by tapping into untraded interdependencies. In Saskatoon, while there is some evidence of this happening around publicly provided infrastructure, there is limited evidence of any existing or developing critical mass of other specialized industry that strengthens the bond between the canola research industry and the location.

A number of public investments in Saskatoon support the industry, including Innovation Place Research Park, SRC facilities and the POS Pilot Plant, which undertakes scale-up work on oil and meal properties for companies with new varieties and consults on oil-processing technologies. As discussed in Chapter 4, this support generally extends to financing. While the availability of capital was central in the innovation process, only belatedly has it generated the human and institutional support mechanisms (e.g. specialized accountants and intellectual property lawyers) to advance the sector. At the peak of the canola development, there is little evidence that there were sophisticated support services for the sector that operated separately or bridged between key institutions.

Moving downstream from the research stage, there were a number of vital forward linkages that increased Saskatoon's attractiveness. Both the regulatory system and market responsiveness matter. As discussed in Chapter 4, the Canadian system has effectively moved these new technologies to the market.

The extensive cooperative farm service networks have been important actors in the creation of a large, receptive and relatively sophisticated farm market for new seeds in the prairies. The Prairie Pools (now privatized and divided among a number of commercial firms), in particular, with an historical delivery share of about 60% and a membership including the majority of farmers in the West, aggressively positioned their organizations as wholesalers for new varieties, partnering extensively with private seed companies to deliver proprietary seeds to farmer members. In addition, the extensive network of not-for-profit producer and industry organizations such as the CCC and the various provincial growers associations ensured the rapid and efficient adoption of the technology.

Beyond the wholesale system, however, the market opens up to the rest of the world, with little evidence that Saskatoon, or for that matter Canada, dominates the system. The 20% of the product consumed within Canada is marketed through the Winnipeg Commodity Exchange or through a number of proprietary processing and retailing supply chains, with little backwards links to the R&D community in Saskatoon. Offshore, Canada's influence was then largely limited to controlling the use of the trademark *canola* on the 80% of the Canada's volume that then made up about half of the world trade in the product. Foreign-owned companies produced all of the resulting processed products with proprietary technologies.

While linkages in the knowledge creation system were the base for this innovation cluster, the downstream capacity to commercialize the product was quite extensive and one of the key factors contributing to the location and expansion of this activity in Saskatoon and western Canada.

In the Saskatoon case, the innovation cluster found a niche as the developer and early adopter of new transgenic canola varieties. The core public laboratories and sophisticated producer and marketing organizations were at the heart of the cluster. It is clear from the evidence, however, that the innovation cluster was never in any way independent or self-sufficient. Rather, it drew heavily on global R&D, worldwide labour markets and speciality services from across North America, with the result that less than half of the value added to the product was captured locally. In short, the system operated remarkably similarly to a traditional trade *entrepôt*.

This alternative model of innovation poses some serious challenges for innovation and development policy. Government policies in this area commonly have a strong mercantilist orientation, with a focus on generating exports while impeding imports. Governments at all levels in many countries actively use their tax and fiscal policy to encourage greater local R&D or to attract global firms to relocate their R&D programmes into their jurisdiction in an effort to generate higher-value exports or to replace imports. This often involves preferential support for national champions or exclusive deals to encourage an MNE to relocate their activities. Usually governments do this without any consideration of the corresponding relationships and interactions that knowledge-based firms require to succeed. If innovation can be thought of as limited to within a firm or within a regional or national community, then such a narrow approach might have some chance of succeeding. But if innovation is truly global, as appears to be the case in many of the life sciences, then narrow, mechanistic, self-sufficiency strategies may either simply fail or at times prove to be counterproductive (as already noted, Landry, 2004, found similar results in the context of the orientation of manufacturing firms and their capacity to innovate).

The evidence from the analysis in Saskatoon suggests that innovation in the agri-food space may be truly global, which goes a long way to explaining why both firms and skilled employees expressed more interest in the innovation community than in fiscal incentives, public infrastructure or government supports. By extension, a mercantilist policy that discourages global links could not only fail to attract but could ultimately drive out firms or researchers as they seek access to the global community. In short, regional innovation systems, like clusters, are very attractive local economic development tools, but they must be nurtured with an appreciation for their partial and incomplete nature. Fundamentally, they are part of a global system and cannot thrive if cut off from the lifeblood of the system – ideas, skilled labour and collaborative platforms.

TRACKING KNOWLEDGE FLOWS

Data from studies in Saskatoon over the past decade can be used to test three hypotheses of knowledge flows. In brief, the analysis of the data for Saskatoon suggests that codified knowledge (STI) flows through global pipelines, while contextual, intangible knowledge (DUI) is embedded in relatively self-contained sector-specific local labour and knowledge markets. While some of the theory asserts that DUI knowledge can link across creative platforms, in Saskatoon these domains tend to remain relatively distinct. Moreover, while recent and planned large investments in soft and hard infrastructure in Saskatoon are clearly part of the story, none of the investments in hard infrastructure unambiguously spurs innovation – soft infrastructure and some indeterminate amount of organically developed social capital appears to be needed to enable these investments to generate innovation. In short, the if-you-build-it-they-will-come mantra that has driven some of the hard investments in Saskatoon and elsewhere is a structurally incomplete model.

One hypothesis is that the economic performance of a community depends on networks to support and sustain local knowledge flows and that physical and knowledge infrastructure can foster the knowledge circulation that underlies these networks. Ultimately, this could involve a mix of people, networks, institutions and infrastructure, but the theory does not make it clear exactly what mix might be optimal.

One place to start with is firms. As a part of a national research project centring on the social dynamics of innovation in the city region, we surveyed 24 firms from recognized and established innovation clusters in Saskatoon – predominantly from the mining and agri-food sectors. Generally, firms credited their competitive advantage to superior innovation, often via collaborations with other players that revolve around large-scale projects executed beyond the firms or organizations involved. While the most common motivation underlying collaboration was the need for efficiency and to cut research costs, a significant number of firms noted their need to stay at the cutting edge of science and technology. Survey results from an analysis of the social dynamics of the biotechnology sector suggest that most companies were highly dependent on face-to-face relationships to keep track of opportunities and threats, many of which appeared to have some local component. In one way, this networking function at least partly addresses the specialization of the community – incoming and outgoing connections are designed to expand the adjacent potentials available for the community.

During the investigation into the social dynamics of innovation at the city-region level, we also examined how firms were connected to innovation efforts in Saskatoon. More than 80% of the firms reported that they had introduced a new product, service or process in the past 3 years – most considered their innovations to be new to the world. Overwhelmingly, the development of these new products, services and processes was the result of internal R&D in reaction to customer and market demand – many firms also cited collaborations that indirectly contributed to the development of the product (e.g. reliability testing of equipment under operating conditions).

Most firms assert that their firm's competitive advantage lies with having the right people on board. This advantage was gained either by building a good team through hiring, by the exceptional skills possessed by the firm's founders or by intensive training. Perhaps most revealing, when firms were asked about their collaborations related to research, development and commercialization, a plurality of firms, largely from the biotech sector, reported that they usually sought help from academics. For many, though, these connections were informal and were often described as simply picking up a phone and calling an acquaintance at the university or elsewhere thought to be able to lend assistance. Only two firms reported dedicated science boards to advise their activities (c.f. Zucker *et al.*, 1998). One particularly innovative firm (in the biotechnology field) reported that they post their problems on a website and offered rewards for possible solutions. In contrast to the biotech sector, mining and software firms tended to turn to consultants, parent or sister companies, or other comparable firms in their sector.

Despite some differences between sectors in Saskatoon, it appears that much of the collaboration and cooperation that facilitates knowledge flows occurs through personal contacts. These collaborations usually involve only brief, informal consultations. While formal connections are still at times important – especially in the context of labour market mobility and relationships with key central actors – the impression from our research is that informal connections between individuals are a more important conduit of information in the average firm's day-to-day operations.

In general, Saskatoon residents were open to exchanging information whenever this was not a direct threat to their company. That the default would be to share knowledge was seen as the natural order of things. Most respondents reported that they had a number of knowledgeable acquaintances outside their company that they could phone in order to seek help on specific issues. Likewise, most respondents stated that they would also probably assist if the roles were reversed. Compensation for these brief consultations was never mentioned except in those instances where the interactions began to run over an extended period. Generally, these consultations were based on pre-existing contacts, but this was not always required. And while these interactions might occur with individuals located anywhere, the majority of the examples given by respondents refered to connections with individuals in Saskatoon.

From this research, one might conclude that most knowledge sharing is done within a framework of social norms instead of market norms. The main difference is that interactions regulated by market norms require (usually immediate) compensation, while those regulated by social norms do not. These concepts represent two fundamentally different ways in which individuals think about social interaction; whether an interaction is perceived as being regulated by market norms or social norms radically changes the nature and outcomes of that interaction. Much research has, rather counterintuitively, shown that the expectation of payment can radically decrease an individual's willingness to render a service to another individual in certain situations (e.g. Fukuyama, 1996; Godelier, 1999; Benkler, 2006). Because of this, Saskatoon's environment of informal connections based on social norms may facilitate a greater level of knowledge transfer and willingness to assist other firms and individuals than other centres where paid consultations are the norm. Additionally, in environments where social norms predominate, reputation becomes more important. It is possible that

Saskatoon's small size and relatively specialized sectoral and labour markets contribute to its social norms-dominated environment by allowing easier tracking of reputational factors.

THE ROLE OF PUBLIC INFRASTRUCTURE IN INNOVATION SYSTEMS

Policy advisors find it difficult to influence firm or individual activities directly, so they often target the use of public investment in a mix of highly specialized hard infrastructure and soft infrastructure to nurture creativity and innovation. As discussed in Chapter 4, the city has been the focus of an aggressive programme of investment in both hard and soft infrastructure.

Investment in infrastructure is undoubtedly an important and vital part of any industrial economy (Phillips and Webb, 2013a,b,c,d). The theory and evidence, however, does not provide any definitive direction to its role and impact on innovation. In the context of Saskatoon, we can see the value of specialized infrastructure as a necessary but not sufficient condition for innovation in the agricultural-biotechnology sector (but not as clearly in the mining sector), while there is some evidence that soft infrastructure, especially the specialized kind, may be critical to realizing the local buzz that makes innovation happen. The emphasis on specialized soft infrastructure seems to indicate that the hypothesis that sectoral diversity drives innovation is not highly supported in the Saskatoon case. While diversity may have micro- and macro-benefits, at the meso-level, it is not clear that it has any particular effect, at least in the case of the Saskatoon economy where the mining and agricultural-biotechnology firms use global knowledge and competencies to compete in global markets. In the context of this research on Saskatoon, we can see that innovation both within specific clusters and in the broader economy is supported by a general focus on innovation and by the capacity to leverage capital in support of foundational infrastructure generated. Much of this capacity seems to be narrowly focused in specific clusters rather than widely distributed, which may simply be the result of a relatively small community. Given that Saskatoon's clusters are relatively smaller in both absolute and relative terms than many others in Canada, it is possible that the local economy has not fully exploited economies of scale in each of these economic areas – hence, there are likely to be ample opportunities for innovation and growth within each of the core clusters, reducing the incentive to seek more diverse collaborations.

Meanwhile, organically generated social capital – in institutions, in policy systems and in the community of creative individuals who propel firms, government agencies, industry associations and various collective efforts – is vital to the Saskatoon economy. It is the medium that generates untraded interdependencies that spurs innovation and development. As noted above, most knowledge exchanges are socially mediated rather than market based. Nevertheless, the strong theoretical focus on meso-level interactions was not witnessed in Saskatoon. Rather, social capital appears to be most actively focused at the micro- and macro-levels, targeted at specific initiatives that would spur development and innovation. There was limited evidence that social capital has facilitated inter-sectoral exchanges that spur innovation in adjacent or tangential sectors. It is unclear what these results tell us about theory and policy. At one level, as suggested above, they tend to refute the diversity hypothesis, which would imply that governments should continue to focus their efforts to enhance specialization by 'building on strengths'. While this has an intuitive appeal in times of fiscal restraint, it may not be the only option. Rather, this focus on specialization may be a phenomenon of a middle-sized isolated resource community at a specific stage of industrial development. Given that there are no direct comparators for Saskatoon in Canada and that the economic and social landscape in Saskatoon has been rapidly changing over the past decade and is likely to continue for the foreseeable future, this may not be a stable, long-term experience. As the Saskatoon economy and society adapt to recent transitions, we may see the need for and emergence of more meso-level diversity and interactions. As with most highly contextualized policy studies, there are few definitive conclusions and few strong policy recommendations one can make. A balanced portfolio of policies, programmes and capital investments in support of both diversity and specialization may remain the best available approach for most governments seeking to support innovation and development.

The Saskatoon case offers some interesting perspectives on the role of social dynamics, diversity and community infrastructure in mid-sized creative and innovative cities. The data offers some limited (but far from universal) support for the assertion that knowledge flows within sectors and clusters linked to specialized structures and processes matters for innovation. Both employees and firms indicated in a variety of studies reviewed that thick, specialized labour markets, formal knowledge transfer and informal knowledge systems are important for knowledge-intensive economic development – it is less clear that these matter for less knowledge-intensive industries. The data does not, however, offer much support for there being any acknowledged externalities between parallel innovative clusters or between innovative clusters and the rest of the economy – if anything, they seem to be solitudes. Nevertheless, there is some support for the need for global pipelines, if for no other reason than to compensate for a lack of diversity within the local community. Firms (and highly skilled individuals) seem to acknowledge the need to remain globally aware and connected, thereby generating more adjacent potential opportunities.

THE ROLE OF CENTRAL ACTORS IN KNOWLEDGE EXCHANGE

The role of public–private cooperation has been recognized as important in the innovation systems, whether in the stylized world of the triple helix or in the more fluid special processes that permeate innovative milieu. Procyshyn (2004) undertook a study of the Saskatoon agricultural biotechnology system to determine the potential roles and impact of public actors in such systems. She used the tools of social network analysis to quantify the processes in Saskatoon and undertook a categorization of the nature of the system. She then investigated whether public institutions have the potential to proactively select winners in such a system or, if not, whether they could create an incentive in potentially innovative enterprises to differentially self-select from the smorgasbord of publicly provided programmes and services in such a way that improved their likelihood of success.

To start, Procyshyn used the results of a major survey of actors in the Saskatoon cluster to identify potential key public sector actors. Eight key actors, all located on or near the university campus or in the Innovation Place Research Park, were unambiguously identified as core to the past and future of the sector's competitive advantages: the university and its nine colleges, AAFC, NRC-PBI, NRC-IRAP, Ag-West Biotech Inc. (now Ag-West Bio Inc.), CLS; POS Pilot Plant and the SRC. These institutions were selected based on the extensive interactions they had among themselves and with other entities in the city; the majority of respondents reported daily, weekly or monthly relationships, and a tendency to view the interactions as too little rather than too much.

One way of examining the role of public institutions is through the measurement approach offered by Theodorakoupoulou and Kalaitzandonakes (1999), who used the social network analysis concepts of density (defined as the percentage of actual versus potential possible engagements) and centrality (defined as the relative connectedness of individual actors with the rest of the system) to quantify the ecosystem and the role of specific actors (these concepts are discussed in some detail in Chapter 9). In their work, they analysed the macro R&D systems in the US and EU systems, undertaking regression analysis to explain the relative commercial success of GM crops based on the nature of the two systems. In their case, they found the high density in the EU correlated with weak commercialization, compared with the less dense hub-and-spoke system used in the USA.

While Theodorakoupoulou and Kalaitzandonakes only examined formal R&D relationships, Procyshyn expanded the analysis to evaluate five functional aspects of cluster activity – R&D, services, financial exchanges, high-quality personnel and networking. R&D is defined as firms or agencies involved in formal collaborative arrangements and joint ventures in pursuit of a specific research objective; services involved contractual fee-for-service and informal yet significant consultations; financial exchanges involved inter-organizational transfers of financing or capital; high-quality personnel related to guest researchers or specialized personnel exchange, joint publications or exchange of membership on boards or science advisory committees; and networking encompassed all those commercial or social interactions that do not relate to formal exchange of goods, services, capital and labour (Ryan and Phillips, 2002).

These functions provided a basis to compare organizational interaction. The eight core actors were surveyed in March 2003 and asked to identify their relationships with each other, as well as with local companies, research departments, government programmes and so on. In total, a list of 95 organizations (including supporting and related industry actors) was presented to these central actors from which they could identify, qualify and categorize linkages (out of a maximum estimated 120+ actors in the agricultural biotechnology cluster in the Saskatoon region).[1] The 95 organizations and eight core actors within Saskatoon's network were engaged in private (37%), public (41%), collective (11%) and mixed (10%) activities. After collecting data from the core actors, supplemented with data collected through related surveys, it was possible to calculate density and centrality measures for the community and the core actors.

The results of the density analysis indicate that, on average, core actors had an average of 89 linkages each, spread across the five functions analysed. The overall network density across all functions was 15% (712 linkages identified relative to an absolute possibility of 8930), which offers prima facie evidence that the community has a range of special processes that have the potential to underpin and accelerate innovation. As one would expect, the density varied by function, from a low of 1.4% for financial exchanges, 1.8% for exchanges of high-quality personnel, 2% for contractual services and 2.3% for R&D links (which is consistent with a predominantly pre-commercial research based community) to, at the high end, 7.6% for networking, which suggests a quite a highly linked community for industrial policy and promotional efforts but relatively weak financial intermediation.

Theodorakoupoulou and Kalaitzandonakes (1999) found a core network density for R&D activities of 9% for the USA and 59% for the EU. According to their study, core organizations in the US network had linkages with an average of nine other organizations, and in the EU with an average of 37 (compared with the average 85 linkages between Saskatoon's core institutions and other cluster actors). This might reflect a fundamental higher degree of integration in Saskatoon, but it is unclear. The number might be somewhat biased upwards because the Saskatoon analysis defines

1 While AAFC is included in the analysis, concerns regarding commercial confidentiality made it impossible to gain access to their responses for most functions; any subsequent references to AAFC should be treated with caution.

data parameters through five types of functional linkages, rather than simply R&D, so that some of the identified average number of connections are duplicated by organization. For example, NRC-PBI has both networking and high-quality personnel linkages with AAFC and both linkage types are recognized in our model. Meanwhile, because this analysis uses the entire population as the denominator (and Theodorakoupoulou and Kalaitzandonakes, 1999, only used research entities), this tended to bias the results downwards.

The eight central actors were then assessed in the context of the overall system to identify their degree of central placement (Table 6.2). According to the analysis, the University of Saskatchewan, NRC-PBI and NRC-IRAP are central actors in terms of R&D activity. In services and financial, NRC-IRAP is the central actor. SRC and NRC-IRAP are central actors in terms of high-quality personnel and Ag-West Biotech is the central actor for networking activity. Our data indicate a wide variance in ranges between functions, with functional centrality ranging between 0 and 100%. Networking was by far the highest with Ag-West Biotech with linkages to 98% of the identified population. The two core actors providing high-quality personnel, SRC and NRC-PBI, received centrality scores of 27 and 20%, while the core financial actor – NRC-IRAP – had a centrality measure of 38%.

Procyshyn (2004) then developed an innovation index for each of the firms in the Saskatoon community, drawing on the Oslo manual definition of innovation – did the firm offer any new or improved products or services that were new to the firm, Canada or the world or is the firm working to develop one – assigning one point for each type of index. The index, which measures both past innovative capabilities and future potential to innovate, ranged from 0 to 7 for each of the coded firms. In order to compare companies, they

Table 6.2. Centrality of core actors by functional linkage. (From Procyshyn, 2004.) Percentages in bold indicate statistical significance.

Actors	R&D (%)	Service (%)	Financial (%)	HQP (%)	Networking (%)
AAFC	–	–	–	–	**62**
Ag-West Bio	–	–	15.5	4	**98**
CLS	0	1	3	1	4
NRC-IRAP	**29**	**44**	**38**	**25**	**64**
NRC-PBI	**36**	20	13	**20**	65
POS	0	5	NA	5	1
SRC	13	**24**	1	**27**	**56**
University of Saskatchewan	**41**	6	0	5	30

NA, not applicable.

were split into two categories. Innovative firms with a score of 6 or 7 almost always had a product or service that was a world first, while non-innovative (or perhaps more appropriately pre-innovative) firms, with a score between 0 and 5, did not yet have a world first. The breakpoint was defined based on the mode of the sample, with 12 innovative firms and ten non-innovative firms distinguished via the index. Procyshyn attempted to determine whether one could identify firms a priori that are likely to be more innovative based on readily available statistics of firm size, structure and history, but found no meaningful results, concluding that it is highly unlikely that public entities could proactively identify winners. The only externally visible variable where there was any statistically significant difference between innovative and non-innovative firms was in the category of patent ownership – 83% of innovative firms owned patents compared with only 30% of non-innovative firms – which suggests that patents may be a necessary but are unlikely to be a sufficient determinant of commercial success. There is ample evidence that most patents are never commercialized, so the evidence of patent ownership is a poor sorting tool for public agencies.

For comparative purposes, the network densities for innovative and non-innovative companies were developed. While the innovative companies exhibited a higher density (35%) compared with non-innovative companies (28%), the differences were not significantly large to make clear conclusions. To understand further the dynamics, Procyshyn calculated the percentages of connections for each category for each innovative and non-innovative company and then conducted a sign test, a non-parametric, two-tailed test for binomial data, to identify any potential correlations. Only three organizations, each with a single function, had significant connections with innovative firms. First, NRC-IRAP was correlated with innovative firms for the provision of research and development – the programme and its staff is explicitly focused on solving technical problems. Secondly, Ag-West Biotech was significantly connected for financial exchanges with innovative firms – its venture capital pool, largely financed by provincial grants, has undeniably taken significant positions in most of the early-stage, high-risk ventures in Saskatoon over the period under review (Ryan and Smyth, 2010). Finally, NRC-PBI offered differential networking services – the research model during the period under review involved collaborative platform research, where NRC-PBI staff was able to gain insights into all of the competing industrial research programmes and share these insights across the platforms (Phillips and Khachatourians, 2001).

Beyond the eight central public actors, a penumbra of supporting organizations were positively linked to innovative enterprises and had the potential to influence the development of the life sciences cluster. In all, seven of the 15 research organizations identified in Saskatoon were connected to innovative firms, as were five of seven civic associations, five of eight government entities and three of five venture capital companies.

The take-home message from this analysis is that clusters can be homes for innovative firms and public institutions and their programmes and services may be necessary for special processes to emerge, but it is next to impossible to target efforts a priori towards firms likely to be more successful. Even if a public agency offers programmes and services that may be sufficient to spur innovative firms, it is hard to get those firms to reveal their preferences or for the agency to sort and select firms to support differentially.

FIRM STRATEGY

In the same way that the special places approach assumes firms pursuing a product-based strategy gain commercial benefits in the form of a greater market share from co-location in clusters, there has been a tendency to assume that firms pursing a resource-based strategy will gain from the special processes embedded in innovation systems. As discussed in Chapter 5, the externalities perspective on agglomeration emphasizes the importance of location factors and public goods to an improved division of labour (specialization), the creation of spill-overs and an overall robust local innovation system. Proximity is deemed to be vital in enabling access to the benefits associated with location factors (also known as externalities). Furthermore, advantages derived from location factors are considered much more difficult to imitate or recreate because they are rooted in participation by multiple organizations. Hence, from the resource-based view of management, location factors are important resources to be developed, controlled and combined to create novel products, processes and services that confer monopolistic advantages (e.g. first-mover advantages and path dependency). In other words, the externalities perspective is related to the resource-based view of management by their shared focus on a process of Schumpeterian competition.

The proposed relationship between the externalities perspective and the resource-based view of management is represented by several variables. The presence of an externalities perspective is measured by proxies for location factors such as a skilled labour force, specialized infrastructure and tailored support services. Karwandy (2008) tested this assertion using correlation analysis to examine the relationship between the innovativeness of the firm expressed through the resource-based approach (based on Procyshyn's index constructed using the Oslo manual definition of competitiveness) and the four variables representing the externalities perspective of clustering. The responses to each of the independent variables were coded into a dichotomous variable with responses of 'yes' or 'no'. The first variable measures the presence of a skilled labour force by asking each respondent if the local labour force possesses any sort of specialization. This measure identifies a labour force that might drive innovation through competition for individual positions as well as competition for the best employees. Additionally, a specialized labour force is presumably bound by common experiences and a shared understanding of language and symbolism that enable ingenuity and inventiveness. The second variable asks each respondent if the local region possesses any unique assets or capabilities. The objective of this measure is to identify the presence of specialized infrastructure. The presence of infrastructure represents both the physical resource platform and the local capabilities to translate infrastructure into a source of competitive advantage for companies. The third variable is a proxy for local competency in support services and ranks each respondent based on whether they use local specialized service providers. Specialized service providers include lawyers, accountants, technical consultants and other service providers that deliver technical skills or high-level expertise to a firm. These support agents are also thought to be valuable vectors for organizational and process innovations. The final variable tests whether firms use local consultants. Consultants are a subcategory of specialized service provider that presumably provide services closely related to the local area(s) of specialization where innovation is occurring. Consultants are posited to have a direct impact on the innovation activity of interest to a firm managed on resource-based principles.

If the resource-based view of the firm is accurate, the four variables representing the externalities perspective would correlate positively with the innovation index. Given that all of the variables are dichotomous the analysis was done via contingency tables (2 × 2). A χ^2 test statistic was calculated for each of these cross-tabs. The χ^2 statistic compares the proportion of respondents in each category with the proportion expected in each category. As a rule of thumb, the expected cell frequencies should all exceed 5 when analysing 2 × 2 tables. The χ^2 statistic tests the null hypothesis that the two variables are independent.

While special processes (i.e. the externalities perspective) are posited in the business literature to be positively related to the resource-based view of management, because both approaches emphasize the development of resources and capabilities in order to achieve successful performance in the long run, Karwandy's findings do not provide strong evidence in support of this proposition. None of the four proxy measures used to represent the special processes that firms could connect to is demonstrably correlated to the proxy measure for firms pursing a resource-based management strategy (as reflected in the innovation index), which could indicate either that the theory is not valid or that firms in these sectors driven by a resource-based strategy are unclear about their strategic role.

CONCLUSIONS

Saskatoon is now, and for the foreseeable future will remain, an excellent case study for testing the evolution and impact of special processes in local and sectoral innovation systems. As a relatively small, regional city that has emerged from an extended period of economic stasis into a period of dynamic growth and development, the city and its resource-based industry are attempting to sustain and capitalize on a sustained resource price boom and offer an excellent laboratory for analysis and policy experimentation.

Research in Saskatoon on the dynamics of economic activity casts some doubt on the robustness of some of the theory. Both market and social dynamics appear to have a role to play. While innovative institutions designed to capitalize on local resources and market opportunities are important, they do not appear to be sufficient to explain creativity; rather, institutional innovation (especially in the context of forward linkages) appears to be, at best, a coincident and possibly lagging indicator of creativity.

Informal collaboration appears uncharacteristically prevalent in Saskatoon. One possibility is that Saskatoon's rural agrarian background contributes to this attitude. There is a tradition for farmers to come together to do work that each of them would be unable to do on their own – the classic example being barn-raising. Certainly, a large proportion of Saskatoon residents (and based on cursory analysis, many of the community leaders) come from a rural agrarian background. However, there is always the possibility that this supposed ethic of cooperation is an urban legend (or should we say a rural legend). It is not clear from the evidence whether people from an agricultural background now living in Saskatoon are more likely to cooperate without expectations of immediate remuneration. It could simply be that this ethic has become so engrained in Saskatchewan culture in general that there would be no noticeable difference between those from a rural setting and those from an urban background. An alternative hypothesis is that mid-sized centres simply exhibit more communitarian features, at least partly because the property of small numbers (especially when it comes to explicit initiatives) overcomes the problem of free riders when trying to create public and common pool goods (Olson, 1965).

7

Stars and Innovation Athletes: People-based Theories and Models

There is a hot debate in academic and policy circles about the emerging hypothesis that the effective functioning of clusters, communities, supply chains and innovation systems depends fundamentally on a small subset of the population that possesses and uses skills, aptitudes and preferences to effect change. Over the years and in different disciplines, we assign different labels to these individuals, such as leaders, entrepreneurs, the talent, stars and creatives.

The logic is that the most important variable determining innovation is not the inherent economic structure of special places or the organization and management of special process, but rather that these are mere contexts where individuals pursue their passions. In this sense, then, the most important factor in any system is the creative, talented, motivated people who are driven by a mix of personal, social and commercial motives. Florida (2002a,b) has proposed the most comprehensive theory, positing that these creatives live and work in cities endowed by technology and talent, and that their heterodoxy is tolerated by the majority. In that sense, these people, while of the system, often think beyond immediate, day-to-day concerns to envisage how things might be rather than how things are. Ultimately, there is a spark of insight and invention that realizes more adjacent-possible outcomes and expands the possibilities frontier in some way. The special processes literature captures some of the essence of this new world. Gibbons *et al.* (1994) offered the concept of the mode 2 world, with purpose-built, heterogeneous networks that engage people in transdisciplinary efforts to generate new forms of knowledge that would be unlikely to occur in the traditional, disciplinarily based mode 1 world. The difference is that the special processes literature does not formally investigate the role of individuals in the systems, nor does it really examine the motives or desires that animate these individuals.

The purpose of this chapter is to examine the characteristics and features of the city region that prove to be salient to talented and creative workers. In short, we investigate why knowledge workers move, what makes them stay and how they engage to realize their creative interests. We are particularly concerned with identifying how agglomeration of economic activity contributes to or stems from the preferences and choices of special people.

THEORY

Economic agglomerations in city regions are the prime focusing mechanism for the literature that explores special people. There is prima facie evidence from economic geography that suggests something special is going on in these areas.

In the first instance, economic activity is seldom equally distributed around the world – as implied by the theory of comparative advantage, cities, regions and nations have a tendency to specialize in those products where their comparative factor endowments match the comparative factor intensities of production and then trade to realize the gains from trade. Starting as far back as Marshall's (1890) observation about the agglomeration of coal and cotton production in Manchester in the 1800s, there are many studies that show that most industries and sectors are highly agglomerated in selected regions, sometimes based on immobile factors of production (such as rich deposits of natural resources and land) but increasingly based on specialized labour and capital that have been developed by human intervention.

Venables (2006) suggested that scale economies may be a big part of the story. He noted that the value added per employed resource rises as the population of a local economy rises, possibly because larger centres offer bigger, more sophisticated markets that can allow land, labour, capital and ideas to be employed in their best uses. This makes intuitive sense. If a lawyer, for instance, is trained and experienced in intellectual property law for biotechnology, they would probably make more money if they were able to practice solely using their advanced capacity. If the local market is too small to allow them to specialize fully, then they will be forced to offer less-differentiated services that will earn a lower wage. The effect of scale is significant. A recent survey of the literature showed that doubling any city's size will increase productivity by between 3 and 8%. Thus, moving from cities of 50,000 to 200,000 would increase productivity from 9 to 24%. Increasing to 1 million would raise productivity from 15 to 40%. In 2005, more than 1 billion people, or about 16% of the world's population, lived in one of

336 urban agglomerations with populations over 1 million (http://www.mongabay.com). More than 45 of these cities had a population of more than 5 million, which on average generates productivity about 50% higher than in the cities of 50,000.

Steven Johnson (2010), citing Geoffrey West from the Sante Fe Institute, suggested that Kleiber's Law – which posits that in nature the metabolic rate relative to the mass of different sized animals is proportional but sublinear – can be applied to cities but with a twist. In some ways, cities do slow down as they get bigger, in that they are relatively more efficient users of energy (and the length of energy lines and carbon footprints) than smaller centres (following a sublinear 9:10 ratio). But unlike animals, economic activity in cities does not slow down as cities get bigger. The bigger the city, the faster people innovate. All productivity-related numbers increase with size, including education, skilled labour, wages and patents. Agglomerating people appears on the face of it to increase both efficiency and productivity.

None of these observations, however, offers any real insight into the causal pathways of innovation. They simply treat the agglomeration effects as part of the black box. West probably put his finger on the core issue – cities tolerate crazy people – but there is little explanation of how this works.

Jane Jacobs (1969) offered one way to open this literature. She is often recognized as the key source of new ideas about the role of cities in the modern study of economic development. Whereas previous literature framed cities primarily as cost-minimizing entities (especially transportation costs), Jacobs expanded the focus to include a much wider range of purposes and benefits. Large cities are seen as the catalyst of economic development and expansion, which is achieved through innovation where 'new work is added to older work'. In contrast to the neoclassical economic explanation, which talks about the more efficient use of resources in urban settings, it is surprisingly the inefficiency and diversity of cities that create the opportunity for innovation. In the drive for efficiency, a city's resources are diverted away from diversity and towards specialization, thereby narrowing the focus and overlooking possibilities for adding new work to old. It is argued that this process reduces innovation and leads to stagnation. Inefficiency, fragmentation, and diversity, on the other hand, increase the potential to add new work to old and thus are more conducive to the generation of new ideas and economic growth. Within this context, a feedback loop develops as diversity breeds innovation, which breeds further diversity. It is the underpinnings of this feedback loop that is of interest to economists and geographers exploring the forces of agglomeration. Over time, the extensive networks created from this feedback loop lead to substantial information spill-overs frequently referred to as 'Jacobs externalities'. Large diversified urban settings create positive forces for agglomeration. According to this argument, large cities are more attractive locations for the establishment of new firms or industries as they can leverage a greater return on economic knowledge. Large cities may also have an advantage if agglomeration forces are added to monopolistically competitive models, as described by Ottaviano and Thisse (2003). Such models generate home market effects where monopolistically competitive industry tends to gravitate to the larger markets and export to smaller markets. In short, the neoclassical economic argument for city agglomeration is that firms come for the cost efficiencies, and Jacobs' argument is that firms stay for the opportunities.

There is some empirical work suggesting that new high-technology firms do tend to flourish in large diverse urban centres. Leamer and Storper (2001) discovered that larger and more globally linked metropolitan areas are more closely linked with innovative activity and that they attain higher economic growth than the rest of the economy. Henderson *et al.* (1995) found evidence for inter-temporal information spill-overs both within a given industry, which they call Marshall–Arrow–Romer (MAR) externalities, and from overall urban scale and diversity similar to Jacobs. As products and industries mature, production can disperse to specialized centres. The dynamic nature of these externalities, where past agglomeration leads to current agglomeration, suggests a tendency towards renewal, at least in large cities. Nevertheless, the degree to which large cities have an advantage in attracting, supporting and renewing their economic capacity remains a crucial question that we discuss further below.

In practice, the balance of economic forces associated with any particular agglomeration – whether based on a traditional industry or a modern knowledge-based activity – appear to be highly case specific. Nevertheless, there appear to be some important common elements in knowledge-based communities. Indeed, as many studies point out, such entities have a fairly typical life cycle. Munn-Venn and Voyer (2004) identified four stages that bear some similarity to Lundvall's neo-Schumpeterian typology examined in Chapter 3. In the early stage when the focus is on the creation and diffusion of knowledge, there is little direct commercialization but suppliers and skilled labour begin to be attracted to the area. In the growth stage, the focus is on commercialization and there is a rapid influx of new firms that are direct participants as well as demanders of skilled labour. In the mature stage, the influx of new firms and skilled labour slows. Finally, the agglomeration faces either renewal or decline. While some high-profile centres such as Silicon Valley have successfully renewed themselves and repeated the life cycle several times, such renewal is not automatic. The history of Detroit since 1970 offers a cold shower to the myth that renewal is guaranteed.

The local knowledge network and technology platform require closer scrutiny. In addition to codified knowledge that may be associated with scientific research, tacit knowledge, which Marshall earlier referred to as 'mysteries in the air', must be present to make a public resource component to the technology platform. Using the terminology of innovation defined by Malecki (1997) and discussed in Chapter 1, this necessitates a know-who as well as a know-why component to discovery activity, as well as with subsequent application or commercialization activity.

It is predominantly the know-who component of knowledge that gives rise to local knowledge networks. Naturally enough, the local knowledge network generally includes so-called local stars or local champions as well as an extended supporting cast. The network interactions among the local actors have been described in the scholarly jargon as local buzz (Bathelt *et al.*, 2004). Local knowledge networks are often supported by overarching industry or trade organizations. From an economic standpoint, such networks act as conduits for the knowledge externalities or technology spill-overs first identified by Marshall. Knowledge externalities may be such that one discovery is a catalyst for another, that one application is a catalyst for another or that a new application leads to new discoveries that at least partially replenish the technology platform.

It should be emphasized that discovery activity, even at first-class universities and laboratories, frequently may not have a strong know-who component. In such a case, the technology platform will tend to be global rather than local and, consequently, the tendency for commercial activity to be localized may be largely or completely dissipated. Thus, the failure of a centre to evolve should not be taken as evidence of either policy failure or lack of entrepreneurial prowess among local players. Advantages to strongly localized commercial activity simply may not exist.

In cases where a local application or commercialization activity draws on a local technology platform, there is the potential for it to become self-reinforcing, again for Marshallian reasons. Application activity, especially in areas such as biotechnology, tends to require intensive use of specialized skills. Not only does localized commercialization activity tend to strongly attract skilled labour, but also the developing pool of skilled labour acts as a magnet to attract further firms pursuing commercial opportunities. Similarly, the localization of commercial activities may stimulate the entry of input suppliers. Meanwhile, low-cost inputs attributable to economies of scale may in turn stimulate the entry of additional firms aiming at commercialization. Where the research role of one or more leading universities underpins discovery activity, the instruction role often strongly supports momentum by facilitating the formation of a local skilled labour pool.

Ultimately, all these lines of research are attempting to understand the role and effect of untraded interdependencies. These interdependencies are often examined through the lens of the special processes discussed in Chapters 5 and 6. What seems to be missing in this approach is the role of individuals, except as leaders of specific organizations or formal initiatives.

Although most investigations of innovation systems have focused on examining the components of the system (through an institutional review) or outputs (using share analysis of value added or employment), we find increasing interest in examining what happens inside the black box of the system. This approach has tended to start with increasingly sophisticated surveys of the actors in and outside a system. The survey results have then been used to correlate business practices against outcomes, to determine if and how connections between actors enhance or impede invention and innovation. The approach has yielded several observations about the various roles actors may play.

The untraded interdependencies in these systems are largely governed by two mutually supporting types of actors: centrally located institutions and formal or informal leaders (sometimes called talent or creatives). Leaders are present in the system in different roles and guises. In many instances, centrally located institutional actors provide a platform for exchange. Central actors could be public research laboratories, formal or informal research collaborations, or industry associations. The price of admission and rules of engagement will vary widely, but ultimately these structures thrive and add value if they are able to generate an environment where participants can feel comfortable exchanging knowledge about what they know that works (sometimes more importantly what they don't know or what doesn't work). Many such systems are formalized via legal contracts or partnerships, which make transparent the rules of membership and activity. A number of models have been tried. Australia has Cooperative Research Consortia and the Commonwealth Science and Industrial Research Organization. Canada has created National Centres of Excellence and NRC regional laboratories. The EU has framework collaborations to complement the research facilities operated by the member states. Japan and South Korea have actively encouraged their *keiretsu* and *chaebol* (post-World War II business models defined as horizontally integrated corporate alliances across many sectors) to create collaborative ventures and empower individual action. The USA has legally sanctioned and supported industry Collaborative Research and Development Agreements (CRADAs). Each is designed at least partly to create a platform for exchange, but for the most part the literature does not explore the relationship between the individual and the collective ventures they pursue.

Other systems operate less formally, with collaborative relationships revolving around industry associations or informal research groups. In these cases, the rules are often not formalized, but rather the systems operate on an implicit golden rule of 'Do unto others as you would have them do unto you'. This type of governing system only works where there is a high likelihood that opportunistic action will not lead to large windfall gains and that the community or network will have a sustained life. Essentially, the system operates in the context of a repeated game, where there is a non-trivial probability that participants will need to interact on a sustained basis, so that one-off opportunistic action will seldom compensate for being ostracized and the attendant loss of opportunities and higher operational costs that would follow.

A second group of actors comprises the formal and informal leaders in a community. The position of a leader will vary depending on the nature and stage of development of an agglomeration. The leader could be a research star, a venture capitalist, an angel investor, a big-linker researcher, an industrial actor or a social entrepreneur. Different agglomerations need different leaders at different times. The difficulty is that there is no conclusive way to identify a leader. Each leader exhibits different idiosyncrasies and capacities, which tends to make most analyses highly subjective and value laden. At one level, we are forced to fall back on the psychological, sociological and economic underpinnings to individual and group action. Some individuals will be driven by economic reward. Other actors seek different returns, in the form of reputation, social standing and various other soft benefits.

The focus on the individual rather than the supply chain in some analyses leads to a wider discussion about the role of society in creating the conditions for technology and talent to come together to breed innovations that go on to generate economic and social value. Florida (2002a,b) suggests the key factor is tolerance for all kinds of diversity (e.g. beliefs, values, epistemologies, morals, races, cultures and lifestyle). Innovation requires change, which tautologically requires some antithetic actors or features. Hence, the system must be able to attract, engage and retain a plurality of voices that challenge the norms.

The challenge for analysis is that talent and tolerance seldom map cleanly on to any single community or, within a community, upon any clearly identified supply chain, industrial agglomeration or cluster. Tolerance is more likely to be a reflection of the broader social milieu while creative talent often lives and operates within, among and between communities and clusters. Hence, the methodological challenge.

Although the focus in many supply chains, geopolitical regions, and technology fields is on building creative environments that will produce optimal inventive effort, we lack the tools to understand how to intervene appropriately. Ingenious, creative people and communities arise almost serendipitously, typically outside any formal governance system. Once they emerge, they are often captured by government, industry or civil authorities, either during the inventive stage or, more often, in the gestation period, when ideas are reduced to practice, legal and regulatory compliance is achieved, and the package is readied for commercialization (Phillips, 2007).

THE ROLE OF THE INDIVIDUAL

Special people are the heart and soul of the third and newest contextual theory of economic geography. These special people are not outliers but rather populate the special places and engage in special processes. While the most recent literature focuses on these individuals in the context of their creative contributions to new technologies and the arts, an extensive literature that predates the current discussion examines the role of individuals in the context of economic and institutional change.

From as far back as the classical period, the literature has focused on the individual as leader. In the earliest period, writers focused on examining and understanding what made key specific individuals into great leaders, such as Alexander the Great. In the modern era, the focus has shifted from looking exclusively at individuals as exemplar leaders to investigating the traits and roles that leaders across the economy and society might exhibit, in an effort to identify what makes them tick and act. The primary stream focuses on the attributes and traits that distinguish leaders from non-leaders. This list, which continues to grow, includes adaptability, conscientiousness, dominance, extraversion, integrity, intelligence, openness to experience, persistence, self-confidence and self-efficacy (Zaccaro, 2007). The literature also recognizes that leaders require followers, and that the potential for leaders to arise and succeed depends on the context (Stogdill, 1948; Mann, 1959). In that sense, a number of different types of leadership have been identified, including: (i) situational leadership (Hersey and Blanchard, 1977), where leadership behaviour is a function of both the attributes of the leader and the characteristics of followers; (ii) functional leadership (Hackman and Walton, 1986), where the leader identifies a need in an organization and personally fills it; (iii) transactional leadership (Burns, 1978), where the group follows the leader to a specific goal in exchange for something; and (iv) transformational leadership (Burns, 1978), where the leader frames and focuses the group on a vision of the future. While much of the literature looks at individual leaders and their roles in hierarchies, especially in politics, the military and formally structured firms, many of these concepts are relevant for less organized contexts. Individuals engaged in a wide range of loosely organized economic, social and cultural endeavours also exhibit leadership attributes.

The second major framing of the individual was as an economic innovator, often called the entrepreneur. Adam Smith, in *An Inquiry into the Nature and Causes of the Wealth of Nations* (1776), asserted that two key concepts unleashed the creative forces of individuals: the division of labour, where individuals would invest their time and energies in those activities for which they were best suited; and then the ability to transact and trade the fruits of their labour. His example of the production of the pin amply identifies why the division and specialization of labour is so powerful. Ten indifferently skilled individuals, he argues, could not produce a dozen pins a day among themselves, but if they organize into a production system, they could produce more than 48,000 pins in the same time. The insight and drive to first break the job of producing pins into small, discrete steps that can be incrementally improved and then to integrate those steps to produce the large number of pins is the first element of the entrepreneurial process. The ability to sustain such gains in productivity then depends crucially on the general 'propensity to truck, barter, and exchange one thing for another', which distributes the surplus production to those who most value it. Smith suggested then that it is a logical and natural extension that 'in consequence of the division of labour, the whole of every man's attention comes naturally to be directed towards some one very simple object. It is naturally to be expected, therefore, that someone or other of those who are employed in each particular branch of labour should soon find out easier and readier methods of performing their own particular work, wherever the nature of it admits of such improvement.' Hence, the division of labour and voluntary exchange bring out the inventive and entrepreneurial spirits.

Ultimately, Smith's framing of individual creativity was encompassed and incorporated into the concept of entrepreneurship, where the entrepreneur sees opportunities to realize adjacent potential opportunities that more effectively use resources. While Schumpeter (1939, 1954) is generally acknowledged as the most important scholar for the modern study of entrepreneurship, he formally acknowledges that, in the 1800s, economist Richard Cantillon probably coined the term and that Jean-Baptiste Say provided much of the insight that is embedded in our modern concept. Ultimately the modern entrepreneur is someone who sees beyond the here and now, envisions an adjacent potential future and uses leadership skills to marshal the resources in pursuit of the opportunity. If successful, the entrepreneur converts their idea into a successful innovation, which, as already noted, Schumpeter defined as the introduction of a new good or a new quality of an existing good, a new method of production, the opening of a new market and the introduction of a new supply of inputs to a production system or a new organizational structure in an industry. While the concept is fundamentally about a process (introduction) by which something new is initiated or adopted, most importantly the entire concept is centred on the key individual – the entrepreneur. John Maynard Keynes (1936, pp. 161–162) asserted that these individuals are not exactly *homo economus*, but rather 'the characteristic of human nature [is] that a large proportion of our positive activities depend on spontaneous optimism rather than mathematical expectations, whether moral or hedonistic or economic. Most, probably, of our decisions to do something positive, the full consequences of which will be drawn out over many days to come, can only be taken as the result of animal spirits – a spontaneous urge to action rather than inaction, and not as the outcome of a weighted average of quantitative benefits multiplied by quantitative probabilities.'

There has been significant effort to figure out who are entrepreneurs, what can be done to make more of them, when and where they will become engaged and how they interact with each other and with the rest of the economy. While Audretsch (2002) and others have done significant research on the role of entrepreneurs generally, their work does not directly explain the agglomerations that seem to be precipitated by entrepreneurs. Similarly, in a less scholarly way, Gladwell (2008), harkening back to the leadership literature, posits that the ICT revolution and many other transformative economic events over the past few centuries were simply the result of the opportune convergence of a primed and ready group of leader entrepreneurs (i.e. those who had put in their 10,000 hours of apprenticeship) and the right context (i.e. the emergence of the assets that could be assembled into what became the ICT revolution). The right people in the wrong place or the wrong people in the right place, and nothing would have emerged. Neither line of investigation does much to explain the localizing effect of entrepreneurial effort.

In the context of our interest in economic agglomerations, the work of Zucker *et al.* (1998) on the role of creative and entrepreneurial scientific stars offers some real insights, and a bridge to the contemporary literature. In 1988, Zucker and Brewer began a study at the University of California, Los Angeles, of elite scientists and their role in founding biotech firms and the effect of commercial involvement on the productivity of those bioscientists. They defined star scientists for their project as those who by 1990 had recorded more than 40 genetic-sequence discoveries or had authored at least 20 articles reporting such discoveries. They identified 337 leading researchers whom they termed 'stars'. These stars were listed as authors on 4315 distinct articles in major journals, which were co-written with another 7718 collaborator scientists who had not themselves met the star criteria. They then examined the role of human capital in the birth of US biotechnology enterprises by looking for the relationship between the location of research stars and the creation of new firms – they hypothesized that successful research scientists would build up personal human capital that is measurable via their patents and publications and then at about the age of 50 would begin to crystalize and monetize their stock of capital by engaging in new start-ups as investors, through the transfer of patented technologies or as advisors (on science advisory boards) or as participants (e.g. as Chief Scientists). They tested this assertion econometrically, concluding that the presence of active stars in a region was strongly positively correlated with the start-up of new ventures, stating that 'at least for this high-technology industry, the growth and location of intellectual human capital was the principal determinant of the growth and location of the industry itself' (Zucker *et al.*, 1998, p. 302). This set of work offers tantalizing hope that individual animal spirits can be tamed and directed to social and economic good. But how one assembles the requisite number and then motivates entrepreneurial stars remained untested.

A number of scholars and practitioners have attempted to crack the chicken-and-egg problem of how to create the critical mass of stars that can spark transformative technological change. The evidence suggests that great institutions attract research stars, which disproportionately win competitive grants that deliver world-class science that is translated to practice through patents and publications. The challenge is to figure out which comes first. Richard Florida in *The Rise of the Creative Class* (2002b) posited that the global economy has undergone a fundamental shift – from the industrial to the creative economy – which has led to the creation of a creative class that drives Promethean-like economic development whose gales of destructive new ideas create permanent change. Individuals within the creative class share four characteristics: (i) they can be highly educated professionals in creative occupations or informal bohemian creatives; (ii) they are highly mobile, not only within their own sector but across sectors, which leads them to make many connections through which information could flow or plans for a collaborative venture could be advanced; (iii) they have a preference for participatory oriented activities (i.e. instead of watching they prefer to join in or to manage the activity); and (iv) they have a desire for quasi-anonymity (i.e. they prefer weak ties such as casual acquaintances to strong ties through formal membership in organizations). Florida hypothesizes that the creative class prefers to co-locate in communities that have an attractive mix of technology, complementary talent and toleration for diversity. His three-Ts hypothesis is quite attractive to most mayors, university presidents and regional politicians because it implies that judicious investments can create a virtuous cycle where creatives will willingly move to join the experience. The difficulty is that, while there is a correlation between the presence of a creative class and economic vitality (Florida, 2002a,b; Gertler *et al.*, 2002), it is not clear what is the cause and what is the effect. Does an attractive and tolerant community spur investment in technology and migration of talent, or is it the other way around? Correlations in and of themselves cannot determine the true causal pathway.

While much of the literature implies otherwise, not everyone in the creative class is focused on the market or artisanal pursuits. Social entrepreneurs are viewed as one increasingly important subset of the creative class and posited by some to be at the heart of the innovation story. The traditional academic view, exemplified by Mills (1956), is that a 'triangle of power' between the business world, government and the military (what some have called the military–industrial complex) held enormous sway and that individuals with money, power and prestige were all-powerful. This has since been broadened by Charles Leadbeater to include a range of social entrepreneurs whose output is social (health, welfare, culture and wellbeing). He characterized these individuals as entrepreneurial, innovative, transformatory, charismatic, people-oriented, visionary opportunists and informal. These people (consistent with the leadership literature) mobilize people and resources to 'regenerate the locality, estate or neighbourhood in which they are based'.

A key question that is actively debated is whether the individuals with these attributes and aspirations – be they leaders, entrepreneurs, stars, creatives or social entrepreneurs – are born or can be nurtured. And, more importantly, what conditions are necessary or sufficient for their efforts to realize economic or social progress?

SOCIAL CAPITAL AT THE CITY-REGION LEVEL

Key creative people are not atomistically engaged in solitary pursuits but were fundamentally interconnected in a variety of social, economic, political, cultural, familial and professional ventures. As the poet John Donne asserted, 'No man is an island entire of itself; every man is a piece of the continent, a part of the main'. At root, it is their interconnectivity that is most interesting.

A review of common definitions on social capital can be classified into four main typologies based on the focus. The first approach asserts that social capital can be understood through action – it facilitates collaborative, cooperative or common actions, or frames expectations of action in a group or a network (Coleman, 1990). Putnam (2000) posited that the most important outcome of civil authorities is that they generate two important types of social capital: (i) bridging social capital derived from mass movements and broad-based service clubs that are outward-looking and encompass people from a wide range of classes and communities; and (ii) bonding social capital from fraternal organizations and exclusive clubs that look inward and reinforce identities. He likened bridging social capital to a sociological WD-40™ and bonding social capital to a sociological Super Glue™. Both are important. Bridging social capital is vital for generating a language, epistemology and structure for understanding and examining potential transformative innovations, while bonding capital (frequently generated inside firms, government departments and agencies or close-knit purposive communities, such as in research teams) is critical for focusing energies on developing, delivering and managing specific, applied innovations.

There are three other conceptions of social capital. Some assert that it is based on the structural or physical position or placement of an individual in a network or structure, positing that well-connected individuals in a group or network can be an asset, allow access to resources, facilitate pursuit of interests and positively affect relationships (Burt, 1992, 2000). Others assert that the psychological placement of an individual in a network or structure can provide preferential treatment, give access to resources and be converted to economic or monetary outcomes (Bourdieu, 1985). Lastly, some define social capital based on resource availability and access (Boxman, 1991) where a person's personal network offers access to a range of resources. Knoke (1999, p. 18) suggested that 'social capital is a process by which social actors create and mobilize their network connections within and between organizations to gain access to other social actors' resources'.

Social capital is both personal and social, but it is fundamentally based in networks of relationships. Only examining networks of relationships would ignore the fact that social capital can never be entirely independent of individual actors and the human capital that they possess. Bourdieu offered three main reasons that the individual level is an important element of social capital. First, individual actors make up the network of relationships; without the individual nodes in a network, no network exists. Furthermore, every actor within a network has a unique set of relationships – relationships in primary, secondary and tertiary forms of association. These unique relationship types result in some actors being more connected than others (i.e. having a higher density of connections), as well as others playing important bridging roles between individuals and associations that would otherwise have no connection. Secondly, structural positions of power – both as a position within a formal organization that is authorized to use and access the concentrated social capital of those individuals that make up the organization and as a reputational position within a network of social relationships – do matter. Thirdly, each node, or individual, within a network of relationships, has a set of attributes beyond their density of connections or positions of structural power; the personality traits, knowledge and abilities of an individual are inextricably linked to social capital.

It should be noted, however, that, while social capital is essential in facilitating the functioning of collaborative network models, the relationship is by no means simple; special people are both the product of their networks and vice versa. There is the potential for a virtuous cycle. Both build on the success and continued functioning of the other. However, in a society lacking the necessary social capital to attract and mobilize the special people who make things happen, it is unclear what can be done.

FIRM STRATEGY AND THE TERRITORIAL PERSPECTIVE

While social entrepreneurs may be important actors in a vibrant social economy, economic progress and the related agglomerations almost all involve, at one time, firms. One gets the impression from some of the literature (e.g. Florida's creative class and the social entrepreneurship literature) that the firm is not the primary focus of these people's efforts. Nevertheless, even the most non-economically attuned individuals often find that the market is one way to advance their passions, values and interests. So a legitimate question is how these special people relate to firms.

The element that connects theories of special people to firms and ultimately economic agglomeration is knowledge. With innovation becoming a focus of firm performance and strategy, the concept of knowledge has risen in importance. Both the product-based view and resource-based view of management have attempted to incorporate the concept into their explanations of firm management. In the case of the product-based view, the unique characteristics of knowledge (non-rivalry and non-excludability) are seen as valuable factors in setting the rules of engagement. From the resource-based view, knowledge is considered the most important resource for generating market value and the primary source of Ricardian rents (Grant, 2002). This raises some question as to whether or not knowledge-based arguments are an extension of the existing product-based or resource-based management styles or if they represent an entirely new perspective on management of the firm.

Grant (2002, p. 135) argued that the knowledge-based view is not a theory of the firm but rather a collection of 'ideas about the existence and nature of the firm that emphasize the role of knowledge'. This argument is accurate in so far as it draws attention to the fact that the knowledge-based perspective is much less developed than the competing perspectives. However, it fails to acknowledge the fundamentally different assumptions that drive the knowledge-based perspective compared with either the product-based view or the resource-based view. For instance, in the product-based view, firm strategic decisions are driven by an external focus where the firm asks how it stands relative to its competitors and how it can position itself to be most successful. From the resource-based perspective firm strategy is driven by an internal focus where the objective is not simply to outperform the competitor but rather to improve continuously at the personal mission of engineering novel products, services and processes valued by the marketplace. In contrast to both of these perspectives, the knowledge-based view of firm management does *not* assume either an internal or external focus as the driver of strategic decisions. Rather, knowledge-based arguments begin with a focus on expanding the knowledge frontier so as to drive the creation of wealth. Even though these efforts to create knowledge can draw the firm's focus into external and internal aspects of performance, the firm continues to make its strategic choices according to knowledge requirements and not competitive positioning or product or service development. For this reason, we suggest treating the knowledge-based view of the firm as a third perspective on firm management.

Before expounding on the strategic implications of knowledge, it is necessary to review the characteristics of knowledge that make it the most valuable factor of production. As discussed earlier, knowledge is non-rivalrous so that once it has been created it can be reproduced at virtually no additional cost. Knowledge is thus a virtually infinite resource once it is created. This resource becomes especially valuable when the laws of scale are applied. The low (or zero) marginal cost of reproducing knowledge means that increasing returns to scale apply to its production. In addition to being non-rivalrous, knowledge is seldom entirely excludable. This means that it is difficult to stop others from accessing knowledge once it has been created. Knowledge is thus valued by society for its spill-over effect. Once created, it is inevitable that the benefits from new knowledge will spread. Strategically, this poses a problem for firms trying to maintain any advantage based on knowledge. Firms must find ways to keep their competitors from accessing and reproducing their knowledge resource. The non-excludability of knowledge is thus believed to act as a disincentive to firm investment in knowledge development.

The conflict between non-rivalry (the vast potential market value of new knowledge) and non-excludability (the risk of not being able to capture that value) introduces what Grant (2002, p. 136) referred to as a 'dichotomy of knowledge-based activity in the economy'. There are activities related to increasing the stock of knowledge (March, 1991, refers to this as exploration) and activities related to deploying knowledge in the form of goods and services (March, 1991, refers to this as exploitation). Exploration (knowledge creation) is a human process and, as such, subject to the limits of human performance. In order to maximize exploration efficiency, individuals specialize. In practice, the exploration process is most commonly associated with scientific research and experimentation activities. Exploitation, on the other hand, requires a diversity of knowledge because products and services are comprised of many pieces of knowledge. In practice, the integration of knowledge can be thought of as the developmental and deployment activities that turn the results of basic and applied research into a saleable or exchangeable item.

Based on this dichotomy of knowledge-based activities, Grant (2002, p. 136) argued that firms competing from a knowledge-based strategy rely on 'some process of knowledge integration that permits individuals to apply their specialized knowledge to the production of goods and services while preserving efficiencies of specialization in knowledge acquisition'. In other words, research-intensive firms rely on strategy to strike a balance between expanding the knowledge frontier and being rewarded for doing so.

With this argument, Grant is essentially expanding the dichotomy of knowledge-based activity to include integration as well as exploration and exploitation. To better illustrate this third dimension of knowledge-based activity Grant (2002, pp. 138–139) posited that there are four important integration mechanisms that minimize the costs of communication and learning: (i) rules and directives (e.g. plans, schedules, forecasts, rules, policies and procedures, standardized information and communication systems) offer impersonal approaches to coordination; (ii) time-patterned sequences help focus each specialist's input independently during an assigned time slot; (iii) routines offer complex yet repeatable patterns of behaviour that function in an automatic fashion in response to a small number of signals or choices; and (iv) high-interaction, non-standardized group coordination mechanisms are often used when there is a high degree of task complexity and task uncertainty (e.g. meetings).

The successful use of these integration mechanisms requires that the individuals participating have some form of common knowledge between them. As Brown and Duguid (2000) observed, it can be difficult to get knowledge to move unless people share the relevant skill, craft or practice. Common knowledge comes in the form of shared language, symbolic communication (e.g. literacy and numeracy), specialized concepts (jargon), shared meaning (metaphor, analogy and storytelling) and recognition of individual knowledge domains.

To illustrate how commonalities enable knowledge to flow, Brown and Duguid (2000) drew a contrast between communities of practice and networks of practice. Communities of practice are small, tight-knit groups of people who work together (e.g. teams of scientists). By virtue of frequent interaction, direct communication, collaboration and the sharing of insights and judgements, these individuals find it relatively easy to circulate new ideas within the group. However, getting ideas out of the community can be substantially more difficult because outsiders do not understand the shared practice. For this reason, Brown and Duguid (2000, p. 29) argued that firms exist to provide 'formal links, joining diverse communities into a coupled system for getting work done and, in particular, for promoting new ideas into marketable products or services'.

Networks of practice, on the other hand, are groups of individuals that engage in the same or very similar practice but do not necessarily work together (e.g. a professional association of accountants). The common practice of these individuals confers a degree of insight and implicit understanding that enables knowledge and ideas to circulate. In networks of practice, ideas are transmitted via indirect communications (e.g. professional newsletters, journals and conferences). The most important implication of networks of practice is that they become an outlet for knowledge flows when information and ideas cease to flow between communities of practice. The denser these networks are, the more fluidly knowledge can flow. In the case of industrial agglomerations, the density of networks can increase to the point where local networks of practice function much like communities of practice. This essentially means that agglomerations facilitate the integration of knowledge by providing alternative pathways of shared practice along which ideas can flow.

While integration mechanisms and the concept of common knowledge are important for operationalizing the knowledge-based view of strategic management, they also illustrate how the knowledge-based view tends to overlap with strategic tactics and practices associated with other views on management. For instance, the impersonal approach to coordination associated with rules and directives (e.g. schedules, forecasts and standardized information) could easily be construed as a product-based response to managing the external environment. Similarly, the sequenced input of different specialists is consistent with a resourced-based response to producing new products and services by developing novel combinations of resources and capabilities. This conceptual overlap occurs because knowledge serves functions in all of the management perspectives.

The implication of this overlap is that it can be challenging to distinguish the less-developed knowledge-based view of management from the product-based and resource-based views. This distinction is best made by recognizing the unique strategic implications of knowledge within each perspective. In the product-based view, knowledge is simply one of several factors of production that a firm must seek to control so as to optimize its competitive position. In this view, the many forms of knowledge (e.g. knowledge of a competitor's actions, fluctuations in demand or changes in supply capacity) can be given a value and can be traded like land, labour or capital. From the resource-based perspective, knowledge is like any other resource or capability in that it is only one ingredient in the recipe for competencies that define a firm's competitive advantage. Like the other resources (e.g. financial, physical and human resources), knowledge can be developed over time and reconfigured with other resources to help a firm achieve its changing objectives. From the knowledge-based view, knowledge is elevated beyond the level of common inputs or resources. Knowledge is treated as the catalyst of value creation within the firm. In this role, knowledge can be applied to the external environment and the associated factors of production to affect the competitive position or it can be used to drive the internal development of resources and capabilities. However, it does not matter if knowledge is used externally or internally – the strategic objective is always to create wealth by expanding the knowledge frontier.

The fundamental principle behind the knowledge-based view of firm management is that value is derived from the creation and application of knowledge. Firms competing from this perspective achieve successful performance through a research intensive effort that finds a balance between creating knowledge, integrating it into marketable items and exploiting the value inherent in these offerings. From this perspective, agglomerations are useful for mitigating the negative impacts of the non-excludability of knowledge while optimizing the economies of scale and scope presented by the non-rivalry of knowledge.

The notion of knowledge synthesis is well represented in the territorial perspective of economic agglomerations (e.g. Granovetter, 1985; Maillat, 1995; Lawson and Lorenz, 1999; Malmberg and Maskell, 2002), which posits that proximity leads to a non-economic structure that mobilizes people and builds trust, common values, exchange of information and ultimately a collective development process. The collective learning process focuses on technology, organization and territory, which, taken together, can represent a 'localized initial context without frontier' (Maskell and Kebir, 2005, p. 8). This perspective, rooted in the innovative milieu approach, embraces the importance of networks, trust relations, reciprocal openness, cooperation and collaboration. As Florida (2005, p. 50) suggested, 'creative people cluster ... because of powerful productivity advantages, economies of scale, and knowledge spill-overs such [proximity] brings'.

Gordon and McCann (2000, p. 520) described this perspective, what they call the social network model, as being built on the belief that trust-based relationships lead to different behaviour on the part of individuals or groups of individuals than market-based relationships (e.g. transactions and contracting) or firm-based relationships (e.g. hierarchical organization). Trust-based behaviour enables greater joint risk-taking, flexible organization configuration and pursuit of mutually beneficial goals. The distinguishing feature of the social-network model is that it is applied in situations where neither market price signals nor internal monitoring systems in firms are sufficient to ensure the success of an activity.

According to Maskell and Kebir (2005), technology, organization and territory are brought together by a collective learning process that is bigger than any single participant. The scale of the learning process precludes large corporations from acting as unilateral developers in a particular territory, at least as far as this model is concerned. Ultimately, individuals and their social capital matter as much as any single organization.

METHODS

Those interested in measuring and testing the role of special people in local and regional innovation have applied a range of relative qualitative and quantitative methods, as well as adapted and adopted some of the new number and graph theory that allows for more sophisticated mapping of social systems.

The starting point for this field of study was the inductive, case-based analysis of specific leaders and entrepreneurs. Carlyle and Galton in the 19th century wrote about the talents, skills and physical characteristics of the men who rose to power and of the leadership qualities in the families of powerful men. In the context of innovation, Hughes (1989) carried on the tradition, examining the role of a core group of visionary scientists and inventors – such as Thomas Edison, Elmer Sperry and Hirum Maxim – who applied themselves in a concerted and sustained way in what some have described as 'invention factories' in order to turn out thousands of patents that changed the economic landscape. In the business world, the role of the entrepreneur has been fully examined through thousands of biographies and autobiographies by the best, brightest and most successful innovators of the past century. This has continued down to the present day – you can usually find a couple of biographies of leading economic, political, social or cultural innovators on any non-fiction best-seller list.

The next obvious step towards method is to move from the individual story to either a meta-analysis of a range of innovators or to proactively survey leaders, entrepreneurs, stars, creatives and social entrepreneurs. These surveys tend to be qualitative assays to identify the key traits, attitudes and aptitudes of these special people, first in order to distinguish between special and non-special people and secondly to determine when and how they are effectively engaged. This often involves an array of behaviour surveys and qualitative assessments of the aptitudes and actions of special people. These databases can then be mined using correlation analysis to assess how these factors and events are bundled.

In an effort to test more formally the theory of special people as the driver for innovation, many social scientists have undertaken more intensive quantitative correlation or econometric analysis. Florida and a range of collaborators in a number of countries have developed indices of individual attitudes and then correlated them to community attributes. In the first instance, Florida (2002a,b) defined his creative class as composed of three distinct groups: (i) super-creatives who are fully engaged in the creative process, including scientists, engineers, educators, computer programmers, researchers and an array of arts, design and media workers; (ii) creative professionals who draw on complex bodies of knowledge to solve specific problems, including those working in health care, business and finance, the legal sector and education; (iii) and bohemians, marginalized and impoverished artists, writers, journalists, musicians and actors who pursue unorthodox lifestyles and hold anti-establishment political or social viewpoints. The super-creatives and creative professionals can be identified and counted in the Standard Occupational Classification System. In the USA and most OECD countries, these two groups comprise up to 30% of the total population. Florida then created a Bohemian Index that charts the concentration of working artists, musicians, writers, designers and entertainers across metropolitan areas, measured as a location quotient, which basically compares regional employment with the national norm. He and others have undertaken correlation analysis to show that the creative class is positively correlated to most indicators of both economic prosperity (e.g. per capita GDP, growth in per capita GDP and personal per capita incomes) and innovation (e.g. production of publications, patents and new firms). Florida also developed a 'creativity index' as a tool to describe how members of the creative class are attracted to a city. The index includes four variables: (i) the creative class share of the workforce; (ii) innovation, measured as patents per capita; (iii) high-technology industry, using the Milken Institute's Tech Pole Index; and (iv) diversity, measured by the Gay Index, a proxy for an area's tolerance, which records the relative number of people in the community who are openly gay. Using this index, Florida periodically rates and ranks cities in terms of innovative high-technology centres. Whether the Florida hypothesis is sustained by the evidence is challenged by quite a few scholars using similar methods. Markusen (2006), for example, pointed out that the creative class is also highly educated, so when the level of education is added to a correlation analysis as a discrete, specific factor, in some cases the engagement in creative ventures is no longer significantly positively correlated with the desired measures of output and innovation.

Finally, some scholars have undertaken more challenging econometric analysis to determine whether and how special people contribute to innovation. As discussed, Zucker *et al.* (1998) regressed the stock of US firms using biotechnology against a number of variables that reflected the co-location with research stars. In all specifications of their model, they found the number of active stars strongly positively correlated to the stock of biotech-using firms, always having a stronger and more statistically significant impact than almost any other explanation, including the presence of venture capital firms. They concluded forcefully that the growth and diffusion of human capital was the 'main determinant of where and when the American biotechnology industry developed'.

There are a few weaknesses of most of the empirical work to date. First, much of the data collecting and analysis have been structured in such a way that it is impossible to use the results to refute the theoretically derived hypothesis – the data and analysis amplify and illuminate the scale and scope of special people in the process but do not demonstrate in any clear way whether they are necessary and/or sufficient for place-based innovation to occur. In some ways (such as Florida's work), it is legitimate to argue that the causality may actually be reversed, where special people are a by-product of innovation processes and not the drivers of those processes. Even when causality is directly tested, as in the series of regressions by Zucker *et al.* (1998), the work is testing for an economic motive and framing for individual action. There is ample anecdotal and scholarly evidence that many special people engage in the innovation process for reasons other than economic gain. In this sense, the models may very well be mis-specified and any significant results may be spurious. Further efforts are certainly going to be needed to fully explore the role of special people.

CONCLUSIONS

People matter. But which ones are special? How do they matter? Where are they most important? When are they most effective? The theory offers some interesting propositions but the empirical work to date has not been able to amplify or refute the theory in any particularly useful way. Any policy based on this theory and evidence is risky.

8 Policies, Strategies and Metrics Related to People in Saskatoon

Richard Florida, best known for his conception of the creative class, achieved notoriety with the publication of the book *The Rise of the Creative Class* in 2002. The timing of its publication in 2002 was fortuitous for Florida in that it followed the dot-com boom and urban planners at the time were open to new and fresh ideas to manage rapidly growing urban populations to ensure the economic success of cities. Florida's work struck a chord with contemporary scholars and economic development officers alike, providing an impetus for dispensing with more traditional economic approaches to development and, instead, pushing for a focus on the influence of cultural amenities. Despite the adulations, Florida and his ideas had their share of critics. Malanga (2004) stated that his work 'reads more like a pop cultural and social history of the Internet generation than an economic-development treatise' and that he made dubious leaps in logic drawing conclusions based on very little evidence. That being said, even Malanga admitted:

> [His] ideas offer a way to talk economic-development talk while walking the familiar big-spending walk ... Yes, you can create needed revenue-generating jobs without having to take the unpalatable measures [such as] shrinking government and cutting taxes ... Florida's ideas are breathing new life into an old argument: that taxes, incentives, and business-friendly policies are less important in attracting jobs than social legislation and government-provided amenities.

Florida and his ideas set the pace for looking at clusters and economic development of regions and cities in a new and interesting light: one that focuses on the role of people. The main tenet of Florida's work is that attraction and retention of talented, technology-savvy and influential individuals is central to the growth and success of cities. Florida (2002a,b,c) utilized a number of indices to measure aspects of local culture such as diversity and openness (see also Gertler *et al.*, 2002) that appear to have gained popularity in the academic world. Most importantly, his work prompted economists and urban planners to look more closely at the role of people in economic development, revitalizing discussion that emerged in the 1960s and 1970s led by Jane Jacobs and Lewis Mumford among others. Florida tested to see where and how the attractive power of a certain set of local amenities (those attractive to the Bohemian or creative class) was related to the clustering of talent and technology-based industry. The hypothesis is that building an attractive, tolerant and diverse environment in the region will also create an environment that will attract the type of skilled and educated employees associated with innovative, technology-based industries.

Glaeser (2004), in a critique of Florida's work, showed that the education or human capital variable accounts for the vast majority of the relationship with growth in population in many communities, in contrast to the Bohemian index or agglomeration of artistic types as Florida suggests. Nevertheless, the discussion around the role of local diversity and creativity in relation to human capital and local growth continues to be an interesting topic in the economic development field.

It is important to distinguish Florida from other approaches. For example, the type of culture that Florida attempted to measure is not the same type of culture investigated by Saxenian (1990, 1994). Saxenian emphasized in her accounts of agglomeration in Silicon Valley and Route 128 the roles of entrepreneurialism and decentralized, flexible and specialized manufacturing. The Saxenian culture (as opposed to the Floridian culture) is all about complex supplier and subcontracting relationships, regional institutions like trade associations, specialized consultants and venture capital and a variety of networks that all help to socialize the costs and risks associated with the regional production system. In that sense, it is more about special processes than special people. As discussed in Chapter 6, Saxenian's approach or paradigm was adapted in the Procyshyn and Ryan (2003) analysis of similar network systems, using social network analysis techniques to measure the network structure of these systems. This approach was designed to explain how the structure of a network operates as an invisible highway upon which knowledge and information can flow within a cluster or regional innovation system. While Florida has been accused of merely repackaging neo-liberal development (Hetherington, 2007; Christopherson and Rightor, 2010), he is right to stress the importance of the individual – in a networked society it is increasingly individual actors that drive social, cultural and economic development.

TOLERANCE, TALENT AND TECHNOLOGY IN SASKATOON

There are a number of ways to study talent, be they stars or more subtle movers and shakers. A couple of approaches have been tried in the context of the agri-food sector centred around Saskatoon.

The earliest and perhaps most basic assay was to identify where there were agglomerations of highly educated, skilled and mobilized researchers. Zucker *et al.* (1998) used the notions of research stars (in their case individuals with more than 20 gene patents) and identified that they tended to cluster and be interrelated with new biotechnology firms. Phillips and Khachatourians (2001) did a similar analysis in the canola research space, finding that research stars (in that case based on the number of peer-reviewed and cited publications) tended to congregate into a small number of centres (with Saskatoon at the centre of the network of effort), and that such aggregations were positively valued by firms seeking locations to do discovery or development research and development.

More recently, a group of about 25 scholars affiliated with the Innovation Systems Research Network, a Social Sciences and Humanities Research Council (SSHRC)-funded Major Collaborative Research Initiative in Canada, adapted Florida's approach to investigate innovation in a range of cities across Canada. In Saskatoon, a survey of creative individuals (called the 'Creatives Survey') was undertaken in July and August 2007. Surveys were coded by organization, copied, and distributed to contacts in a wide range of industrial, social, governmental, professional, service and artistic organizations in Saskatoon. These organizational leaders were then asked to distribute the survey to as many individuals as possible in their respective organizations. More than 1000 surveys were distributed and 115 surveys were returned, but only 109 surveys were usable. The survey results that were rejected were either incomplete or were answered from an institutional rather than an individual perspective.

Respondents of the Creatives Survey included people with diverse backgrounds and experiences. Respondents had an average age of 35: 17% were over 50, 24% were in their 40s; 19% in their 30s and 39% in their 20s. Approximately one-fifth of the respondents were born and lived most, if not all, of their lives in Saskatoon; another two-fifths were born in the province and migrated to the city at some point. The remaining 37% came from outside the province: 16% from western Canada; 5% from central or eastern Canada and 16% from abroad. Based on the individual's stated date of arrival in Saskatoon, we were able to calculate that on average the survey respondents had spent 55% of their cumulative living in Saskatoon; only 28 individuals reported having lived all of their life exclusively in and around Saskatoon.

On the face of it, the survey would appear to have targeted and included at least one significant talent pool: namely, those with advanced degrees or high-technology workers. Only about 10% of the respondents had terminal education at the secondary level, 25% had a technical diploma or degree, 36% had an undergraduate degree and 29% had graduate training, including 21 individuals with PhDs or other higher doctoral training. Sixty respondents (more than 50%) got all of or their final education in Saskatoon. The average distance for the 49 who did their final schooling elsewhere was 1325 miles (standard deviation (SD) of 2106) from Saskatoon.

The Creatives Survey demonstrated that the respective talents of these high-technology workers were being applied in the local economy in a variety of ways. Of the 109 respondents, 108 reported on their last five jobs and on their tenure with those jobs. Furthermore, all were employed at the time they were surveyed (this is not surprising, as the survey was administered through work-place institutions) and generally were mobile. On average, they reported 3.4 jobs in the 5-year period, with an average tenure of 3.8 years per job: the time individuals hold their jobs (sometimes called tenure) in the broader Saskatchewan economy averages more than 10 years. About 21 individuals reported average tenure of more than 5 years (this group on average reported three jobs and an average tenure of 9.8 years per job); the other 87 individuals reported that their average job tenure was 2.3 years (based on an average of 3.5 jobs). Given the direct and indirect impacts that these high-technology workers are purported to have on a city region's economy (as seen in the literature and in the preliminary findings of the Creatives Survey), we went on to develop a Talent Index to correlate talent against some of the measures of tolerance and other attractors in the community to determine how they may relate. In essence, the Talent Index, in this case, is used to answer two questions: why do high-technology workers move to Saskatoon? What makes high-technology workers stay?

There is no universally accepted definition of talent. In our study, we constructed a naïve index that attempts to capture elements that may reflect talent. In particular, the Talent Index is a straight average of the percentage of a person's life spent outside of Saskatoon (which ranges from 0 to 98%) and their highest level of education (senior matriculation = 0.2, technology diploma = 0.4, undergraduate degree = 0.6, masters = 0.8 and PhD = 1.0). The Talent Index could range from zero to 1.0 – the average value for the 109 respondents was 0.53 (SD 0.25).

While other investigators that use this type of methodology often chose to focus on the professions or sectors the individuals work in (e.g. Florida, 2002a) or on individual responses to personal questions about attitudes and motivations (e.g. Webb, 2009), we instead focused on the *role* of creative people in industrial settings. In this case, this is where an individual's background education and willingness to match it to new opportunities are used as proxies for creative capacity. Mobility is used as a proxy for talent as it reflects an individual's opportunity cost and willingness to proactively move to match their skills with new opportunities. The further they go, the more opportunity an individual has to access new networks and to participate in and share new experiences. As they combine these new experiences with their background, they cultivate new knowledge and develop new skills with new colleagues and neighbours. While education is not a perfect indicator, higher levels of accomplishment can reasonably be assumed to generate new opportunities for recombining knowledge that will generate inventions and innovation.

One of Florida's fundamental assumptions is that talent is attracted by the presence of technology, other talent and tolerance. Tolerance, by definition, can imply acceptance of diverse views related to race, age, sexual orientation and beliefs. Hence, no single, all-encompassing measure of tolerance can be used. The Creative Survey probed this in several ways.

In the first instance, we queried all respondents as to whether the city was a tolerant or welcoming place (i.e. in terms of race/ethnicity/secularity/general equality in their field). On a 10-point scale, where 1 is intolerant and 10 is maximally tolerant, the average of all responses was 7.4 (SD 1.7). The correlation coefficient between the talent measures for each individual and their response to this tolerance question was 0.07, which is not statistically significant from zero at 99% confidence. Hence the views related to tolerance were not significantly related to the respondent's level of talent.

As a second test of tolerance, we asked the respondents for their racial or ethnic background. The vast majority of the respondents reported that their background as northern European or Caucasian. Ten respondents reported they were members of a visible minority. While 15 individuals reported that they had 'encountered discrimination in Saskatoon in their field', we found no statistically significant evidence that this was related to race (the correlation coefficient between visible minority and discrimination was –0.0347, which was not statistically different from zero). Only one of those reporting personal discrimination was a visible minority (an Aboriginal female). The others who expressed concerns about discrimination described themselves as of European or Caucasian descent, identifying concerns related to race (eight respondents), gender (seven, including one male referencing reverse discrimination), sexual orientation (two) and ageism (one). It is important to note that we are not saying that racial discrimination does not occur in Saskatoon – it has been, and continues to be, an important issue that the city (and all of Canada) needs to address. Levels of income inequality between the national average and First Nations or new immigrants are stark reminders of this.

Another measure of tolerance is whether the city of Saskatoon is an urban centre that is open to experimentation and creativity. Again on a scale of 1 to 10, with 10 being most tolerant, the average response was 6.3 (SD 1.6). We tested this with talent and found no statistically significant correlation between the Talent Index score and the view on this measure of tolerance.

We went on to test which characteristics of living and working in Saskatoon make it an attractive or unattractive place. We presented respondents with 20 specific community features and asked them to rank each on a 5-point scale, from very negative to very positive. We also asked them to rank the top three positive features and the top three negative features. Table 8.1 shows the results. While all but the tax regime generated average positive responses, only four responses were statistically positive – commute time, community environment, suitability for raising children and work environment in the firm – and got the most net positive votes. Conversely, four factors generated the most net negative votes: natural environment or climate, tolerance, salary, and local and provincial tax regimes.

While Table 8.1 is illuminating, it does not directly address the relationship between those characteristics and talent or creativity. Phillips and Webb (2008) also undertook a correlation analysis between the talent index and the individual responses to the attractiveness of the 20 identified characteristics. Only six of the 20 were statistically correlated. Salary, cutting edge work and affordable living were all positively correlated with the Talent Index (with Pearson's correlation coefficients of between 0.21 and 0.25 at the 95 or 99% confidence level), while restaurants or night life and proximity to family and friends were negatively correlated with talent (with Pearson's correlation coefficients of between –0.33 and –0.39 at the 99% confidence level). The positive measures are all about an individual's opportunities and would normally be viewed as part of one's personal cost–benefit calculations of career and location. The positive correlations with talent would tend to suggest that the work environment in Saskatoon is a positive driver for talent attraction and that it is perceived to be appropriately compensated (perhaps both in normative and purchasing power terms). In contrast, Saskatoon is somewhat less attractive from the perspective of some talented individuals because of its limited night life and dislocation from friends and family (who, given the mobility of the study respondents, are often in far-flung parts of the world). This result conforms with earlier work done by Phillips and Khachatourians (2001) and Phillips (2002).

One way to test these results further is to examine whether the local economy supports mobility of knowledge between jobs and sectors. The survey asked respondents whether Saskatoon's economy enables mobility between sectors. On a 10-point scale (1 = none; 10 = high), 58 individuals responded with an average of 6.5 (SD 1.6) that the economy facilitates mobility. When asked whether the respondent uses knowledge gained in other sectors in their current work (0 = never; 10 = frequently), 62 responded giving an average of 6.6 (SD 2.2). While earlier results seemed to demonstrate

Table 8.1. Rating of community characteristics. (From Phillips and Webb, 2008.)

	Mean	SD	Positive votes 1	Positive votes 2	Positive votes 3	Negative votes 1	Negative votes 2	Negative votes 3	Net votes
Commute time	1.33	0.99	9	15	11	1	1	1	+32
Community environment	0.94	0.85	4	4	4	1	1	3	+7
Suitability for raising children	0.93	0.89	8	9	5	0	1	2	+19
Work environment in firm	0.91	0.88	4	6	6	0	1	1	+14
Proximity to family	0.84	1.25	13	7	6	5	6	1	+14
Proximity to friends	0.80	1.05	3	5	9	2	4	4	+7
Recreational/cultural amenities	0.76	0.89	1	6	2	1	6	1	+1
Ability of partner to find work	0.65	0.91	7	2	1	3	3	2	+2
Quality of schools	0.63	0.78	2	3	3	1	0	2	+5
Availability of job opportunities/ advancement	0.61	1.05	6	5	10	3	4	8	+6
Community safety	0.58	1.07	5	5	5	4	3	7	+1
Social and ethnic diversity	0.57	0.83	0	2	5	0	5	2	0
Cutting edge work in the field	0.46	1.08	9	3	2	5	2	2	+5
Affordable living	0.44	1.30	10	4	4	15	5	4	–6
Openness to experimentation and creativity	0.41	0.90	1	1	3	2	2	3	–2
Natural environment/ climate	0.32	1.17	2	4	2	13	6	6	–17
Restaurants/nightlife	0.31	0.88	1	2	2	2	4	4	–5
Tolerance	0.30	0.91	0	1	2	6	4	7	–14
Salary	0.15	1.02	1	2	2	5	10	8	–18
Local and provincial taxes	–0.17	0.96	1	0	2	13	11	6	–27

that high-technology workers enjoyed employment mobility, these responses tend to suggest that mobility (specifically employment mobility and knowledge transfer) may not be overly effective within the Saskatoon city region; there was no significant correlation between the responses and the Talent Index.

From these results, we might infer that Saskatoon might not be overly tolerant of creatives and innovators. Whether this is fundamentally true (and the evidence remains inconclusive), it is still possible to determine whether or not the perceived lack of tolerant, innovative zeal in the city has an effect on the economy. The survey asked respondents whether they had ever started a new business or were employed by a start-up or early-stage business. Twelve respondents reported establishing a business (eight for services, one produced goods and three did not record the nature of the enterprise) and another 21 reported that they had worked for a start-up firm at sometime in their career. When the Talent Index was tested against an index of firm expansion (0 = never connected; 1 = employed or entrepreneur; 2 = employed and entrepreneur), a very small positive correlation was found (0.06) but was statistically insignificant.

Finally, the survey asked what particular aspects of Saskatoon facilitate creativity in the city (Table 8.2). In all, 80 responses cited specific aspects about Saskatchewan that affected creativity. Of those who offered their thoughts, 26 respondents reported that specific institutions (especially industry or infrastructure) facilitated creativity, while 31 reported that cultural aspects of the city supported creativity. Those citing institutions focused on the relatively large role the scientific community plays in the city, mentioning the infrastructure at the University (including the CLS and the federal research laboratories), the biotechnology firms and the nature of a competitive yet cooperative community. Those citing the community and cultural aspects of the city mentioned amenities (e.g. cost and variety), rural,

Table 8.2. Industrial, institutional, community and cultural attributes supporting creativity.

	No. of citations	Specific attributes cited
Industry and institutions	26	Inclusiveness; large scientific community; competition and cooperation Biotech industry Research infrastructure (e.g. university, federal laboratories); large research community relative to small city
Community, culture and cultural amenities	31	Size; amenities; lifestyle; pace; cost; sense of community Cultural events; affordable and accessible activities Rural/agrarian/small town virtues (friendly, accepting, volunteerism)
Yes	5	Positive responses without any specific industry or community feature identified
None	20	Negative features included isolation, conservatism

agrarian, small-town virtues (e.g. friendliness, acceptance and volunteerism) and access to affordable and engaging cultural events and facilities. Twenty-five respondents reported that they thought there might be some positive features, but they could not think of any or reported only negative features (e.g. their perception of Saskatoon as a small, isolated and conservative community).

Statistical tests were done to look for relationships between the citations and the Talent Index. The correlation coefficient between the Talent Index and industry and institutions was 0.298 (statistically significant at the 99% level), indicating that those who have higher talent measures see value generated by those institutional and industrial features that are unique to Saskatoon. No statistical correlation appeared between the Talent Index and community and cultural amenities or the negative responses. This does not necessarily mean that cultural and community attributes are not a contributory factor, simply that they were not differentially recognized by those who formed our talent pool.

When examining the importance of tolerance, social and cultural vibrancy and mobility in attracting, integrating, retaining and motivating high-technology workers in Saskatoon, we found very little to support the literature surrounding place and creative workers. Rather than being attracted to the city region by characteristics of place usually associated with members of the creative class, the most creative individuals in our survey were more likely to be attracted and retained by the professional and commercial prospects of place – reflecting the positive correlation between talent and salary, cutting edge work and affordable living. Ryan *et al.* (2011) explore this facet of city regions in the context of Calgary, the nearest and most important neighbouring city to Saskatoon, using a more quantitative approach in the context of an 'embeddedness' indicator. The authors use the indicator to capture factors reported about innovative workers' decisions both to move to and remain in the Calgary census metropolitan area. The indicator, in question, leverages the Mitchell *et al.* (2001) definition where embeddedness is at the firm level. Using a grounded theory approach, Ryan *et al.* (2011) tested a series of hypotheses around potential attraction and retention factors such as socio-cultural, economic, network factors, and geographical and structural amenities.

Analytical results across both the Calgary and Saskatoon census metropolitan area are, broadly speaking, economic or business-related in nature. For Saskatoon, the results suggest that workers are lured to the city region because it is where the jobs are. However, the ability of the Saskatoon city region to retain these high-technology workers suggests that, once they are captured by the professional and commercial aspects of place, the broader spectrum of community and cultural amenities are either adequate or at least do not fully offset the other benefits. For Calgary city-region innovative workers, both personal and professional networks are perceived as key attraction and retention factors. This echoes the findings of Mitchell *et al.* (2001) where social networks appear to lead to professional benefits. This is also consistent with the literature where Bolan (1997) suggested that locational attachment comes about as a result of switching costs associated with changing social networks (Ryan *et al.*, 2011).

CREATIVITY, LEADERSHIP AND SOCIAL CAPITAL

In 2008, an analysis was done on social entrepreneurs in Saskatoon (Webb, 2009). Four main structures in Saskatoon – the University, government, business and civil society – were examined involving a range of individuals who play key roles in the community. Taken together, a long list of key actors in the Saskatoon city region was compiled and a subgroup was then surveyed to determine the nature of their economic, social and cultural interactions. These links were used to identify different types of network leadership, which were then correlated with a set of personal attributes to determine the motivations and influences on leaders.

Broadly speaking, the analysis suggested that creative social entrepreneurs are not unambiguously the primary creators of social capital in the Saskatoon city region. However, a more thorough analysis of the data offered more nuanced findings. From this analysis two broad findings were highlighted.

First, the data analysis consistently indicated a strong positive relationship between high levels of professional creativity and central placement as leaders, power brokers and gatekeepers (see Chapter 9 for more on these features of social network analysis). Professional creatives often have large far-flung networks, act as connectors by bridging different groups and individuals, and hold positions of formal and informal structural power in their social networks. One can infer from this that individuals who are highly educated, have creative jobs and are relatively highly inter-sectorally mobile are engaged in efforts to positively influence social capital creation at the city-region level.

A second implication of this research is drawn directly from the result that professional creatives influence overall social capital creation in the city of Saskatoon. As Florida (2002a) suggested, creative individuals are drawn to creative cities – places with vibrant and tolerant social and cultural atmospheres. The hypothesis that creatives are attracted to a location because of certain aspects of the place is probably (at the very least anecdotally) true. However, once present in a city, creative individuals – especially professional creatives – can have a significant impact on the maintenance and growth of vibrancy in the city. Professional creatives become involved in the community, they create or join organizations and they can bridge or broker between individuals and institutions. Quite simply, members of the creative class are not just consumers of social and cultural vibrancy, they are also producers.

Thus, what emerges from the literature is a central hypothesis that can be tested: creative social entrepreneurs (individuals with high entrepreneurial and creativity levels, both professional and informal) are the primary creators of social capital (societal, associational and individual) at the city-region level. Webb (2009) broke this down into three narrower assertions to test: (i) creative social entrepreneurs have large, far-flung networks of relationships that span the entire community; (ii) creative social entrepreneurs are the brokers (connecting people and groups) in the network; and (iii) being in a structural position of power in a heterarchical, mode-2-style network is synonymous with being a creative social entrepreneur.

To start, it was necessary to identify and study leaders and creative social entrepreneurs. Scott (2001) suggests that most studies of leading members of business, government and civil society use a mix of reputational, decision-making and structural methodologies. The classic reputational analysis of power structures in city regions was in Floyd Hunter's *Community Power Structures* (1953). Hunter compiled a list of prominent people from four groups assumed to have power connections – businesses, government officials, civic associations and society activities – and then recruited a selection of judges to identify the most influential individuals, who were asked to take part in the detailed study of power in the city region. Scott, (2001, p. 86) cautioned that the reputational analysis methodology 'at best, provides evidence on images of power'. The difficulty is that judges limited their responses to men they believed to be most influential, and not a single woman or African American was on the list compiled from the various databases that were given to the judges (which was a problem given that this study was conducted in Atlanta, a city with a vibrant and substantial African American community) – in short, reputational analysis will only give an image of who has power, which will be tainted by the social and cultural norms that influence those whose opinions are being studied.

Dahl (1961) was largely critical of Hunter's approach and instead proposed a decision-making approach for identifying and assessing influential individuals in a city region. While Dahl again started with a list of influential individuals, he used it to identify individuals who were potential holders of power. Dahl hypothesized that there was an important distinction between influential individuals who had the potential to affect decisions versus those influential individuals that actually did influence the decision-making process. Given the right set of circumstances the decision making approach suggests that any individual or group could have potential influence on the policy proccss – any individual, given the correct variation in resource use could lever their potential power into actual power. Critics of Dahl, such as Bachrach and Baratz (1962), were concerned that the decision-making approach fails to examine decision-making processes that occur behind closed doors that serve to define and limit the agenda. In other words, informal actors often exert deeper, more pervasive, hidden power. In essence, Dahl examined only relational power and not structural power (Strange, 1988).

A third methodological approach for identifying influential individuals – structural power – incorporates the insights and critiques from the other two approaches. A structural power methodology seeks to understand the structures that the decision-making apparatus operates within (i.e. incorporating non-decision-making processes into the study). This enables an observer to learn about perceptions of power and thus incorporate reputational analysis. The starting point of any analysis of individuals of influence must therefore be with a study of structural power. The structural approach posits that, while the identity of particular individuals is important, the attributes of the positions that they occupy are more fundamental to power relations. The paradigmatic structural power study was conducted by Mills in the 1950s and reported in his work *The Power Elites*. This study revealed that, among the power structures in society, there were a class of informal institutionalized power elites. These elites were defined as 'men whose positions enable them to transcend the ordinary environments of ordinary men and women; they are in positions to allow them to make decisions that have major consequences' (Mills, 1956, pp. 3–4).

In analysing the structural position of individuals within networks, one must make concessions for the attributes that are associated with individuals of interest, attributes that can create new qualitative and quantitative insights. Methodologically speaking, Webb (2009) identified three types of attributes as they relate to the character of creative leads: (i) the level of creativity (professional and informal); (ii) entrepreneurial capacity; and (iii) formal community involvement.

The literature on creative individuals outlines many different types of creative leaders. Webb (2009) focused on two creative types: professional creatives (e.g. academics, architects and graphic designers) and informal creatives (e.g. artists, musicians and thespians). While there are differences between these two creative types, they both still reflect Florida's characteristics of members of the creative class (creativity, horizontal hypermobility, participatory activities and quasi-anonymity). Using data gathered in a survey of Creative Social Entrepreneurs, Webb was able to create measures of individual creativity to reflect both professional and informal capacity. The professional creativity measure was created by weighing the level of education, current employment and sectoral mobility in their employment history, while informal creativity was measured using the types and numbers of hobbies of each individual.

Webb (2009) then used the Gallup 'entrepreneur perceiver' questionnaire to identify life themes – 'a consistent and recurring pattern of thought and/or behaviour' – and to develop a questionnaire to assess the attributes that contribute to successful entrepreneurship (Clifton and Harding, 1986, in Bolton and Thompson, 2004). The themes involved a range of perceptions, predispositions, goals and observable responses to life circumstances. These data were used to determine the level of entrepreneurial characteristics that each person exhibited; the more of the 12 themes an individual responded to, the more the individual might exhibit Keynes' 'animal spirits' and engage in entrepreneurial activities.

Finally, the survey probed individuals' formal community involvement by examining the type of associations that an individual was involved in (e.g. community-based organizations, sports teams and political parties) and in what capacity they were involved (i.e. member, active member, or volunteer).

Using data from the survey, Webb (2009) used the framework in Fig. 8.1 to investigate the nature of creative social entrepreneurs in the Saskatoon city region. The framework allowed the examination of both informal and professional levels of creativity, measuring entrepreneurial characteristics and analysing the breadth and strength of formal community involvement,

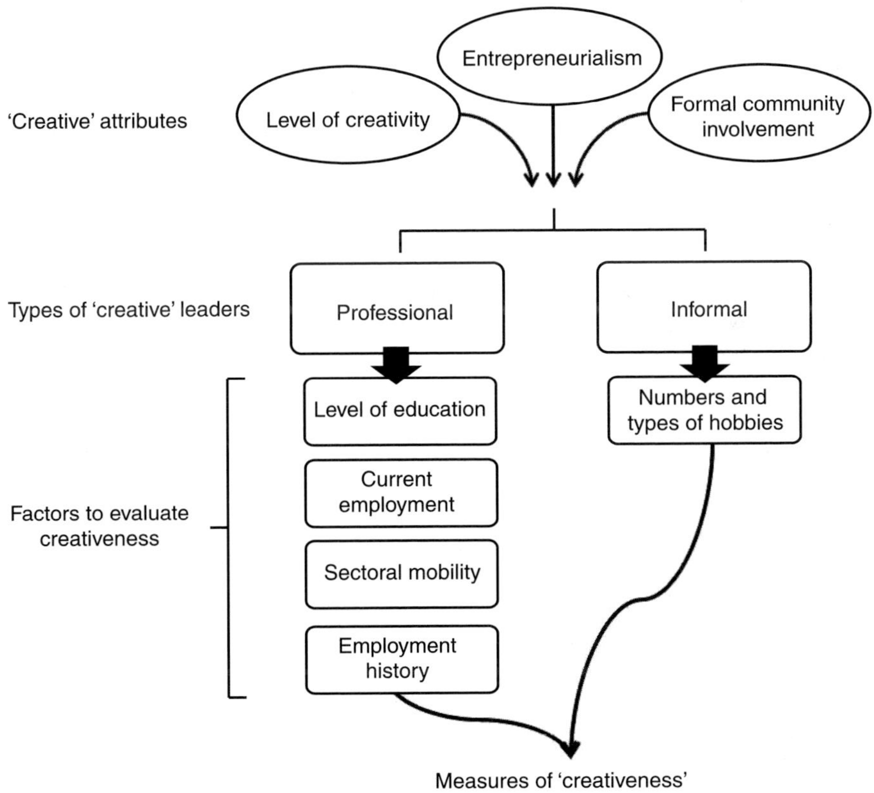

Fig. 8.1. Framework for analysing creative leaders and activity in communities.

bringing into focus the structural position of these individuals in the networks of which they are a part.

Thirty individuals (from a reputational list of 253 leaders) took part in the Webb study. These individuals were from all four major structures in the Saskatoon city region (five in government, nine at the university, seven in industry, nine in civil society and two others; two participants held positions in more than one sector, so while there are only 30 participants, there are 32 sectoral tallies). Nineteen of the participants were female and 11 were male. The ages ranged from 31 to 64, with an average of 48.9 years. Eleven of the respondents were born in Saskatoon and the rest, aside from one individual who did not specify, were born outside the city. Of those individuals who moved to Saskatoon, the average number of years spent living in Saskatoon was 22.8 (with a range of one to 52 years). Webb concluded that the 30 respondents were, on average, in parts of their life cycle where their careers were well established and that they had been located in the city for long enough to create substantial social networks. Beyond the basic biographical data outlining who the respondents were, three important categories of attributes were investigated – creativity, entrepreneurial capacity and level of formal community impact. Webb found that the mean of professional creativity was 0.42 with a SD of 0.15, the mean of informal creativity was 0.23 with a SD of 0.16, the mean of entrepreneurial capacity was 0.81 with a SD of 0.09 and the mean of formal community involvement was 0.17 with a SD of 0.18. While the means for professional creativity, informal creativity and formal community involvement may seem low, the numerical values have little absolute relevance. Rather, these values are important insofar as they provide a relational relevance between respondents; using these values, it is possible to order individuals according to their levels of human capital and whether those who are above average have a greater impact on the governance of the city region.

In the first instance, Webb undertook a correlation analysis to test for any relationships between the three attributes. He found a negative relationship between professional creativity and both informal creativity and formal community involvement, a positive relationship between informal creativity and entrepreneurial capacity and a positive correlation between entrepreneurial capacity and formal community involvement (albeit all below 0.24 and none significant at the 90% level of confidence) (Webb, 2009).

It is important to note, at this point, that there is a division in the literature between Florida's creative class (akin to professional creativity) and Bohemians (informal creativity). While both are creative, they represent distinctive manifestations of creativity. In the context of the Webb study, however, as the small (–0.13) correlation coefficient between professional creativity and informal creativity was not statistically different, it was not possible to prove or disprove this point in the context of evidence in Saskatoon.

Webb also undertook a separate correlation analysis between individuals who are both highly creative and entrepreneurial and their formal community involvement but found only a very small and insignificant positive relationship. This first step in investigating the hypothesis that creative social entrepreneurs are the primary creators of social capital at the city-region level failed to show any significant results in Saskatoon. One of the key characteristics of creative social entrepreneurs that theoretically make them creators of social capital is that they are deemed to exhibit a tendency towards taking part in participatory activities. Creative social entrepreneurs, theoretically, are more likely to join associations – associations that house much of a community's associational social capital. As a result of this finding, Webb undertook a broader investigation of creative social entrepreneurs and social capital creation using social network analysis (see Chapter 9 for more on the methodology and its application). It is worth recalling that Webb made three propositions about creative social entrepreneurs: (i) they have large, far-flung networks that span the entire community; (ii) they are bridgers or brokers, connecting both people and groups; and (iii) they occupy structural power positions in a heterarchical network. Webb used three different social network analysis measures of centrality (closeness, betweenness and eigenvector) to attempt to confirm or reject the statements about creative social entrepreneurs in the Saskatoon city region. Each measure of centrality, developed from the respondent's social mapping of their work ties, community involvement connections, casual acquaintances and friendships, were tested for correlations with their personal attributes (professional creativity, informal creativity and entrepreneurial capacity).

The analysis of these comparisons between social networks (social capital) and individual attributes (human capital) was organized using the associability typology. This typology breaks down the different forms of engagement in informal networks and formal associations into three main associative forms: primary associations, secondary associations, and tertiary associations. Primary associations are most often composed of ties of family or kinship; they are often non-voluntary in nature. They are exclusive, with high barriers to entry, and form a system of mutual obligation. Lastly, primary association ties are strong ties characterized by closeness and a high frequency of contact. Webb chose to use the friendship ties revealed in the social mapping as primary association relationships. While these ties are not non-voluntary, the degree of exclusivity and its expected system of mutual obligation make this a close fit. Primary association ties do form social capital, but this type is by nature more closed then the social capital formed in secondary association ties; while social capital formed in primary associations can be leveraged by members of the group, it is rarely a public good that all of society can make use of. Secondary associations are predominantly composed of community association ties, which are characterized by low barriers to entry, heterarchical modes of interaction (versus the hierarchical mode of interaction that characterize tertiary associational ties) and goals that are variable. Community involvement and casual acquaintance ties revealed in the social network mapping best fit as secondary association ties. The secondary associational ties are probably the most important measures of the level of social capital in a community, as they are the creators and incubators of social capital. Tertiary associations are often composed of an individual's ties to firms. As a result, work-based social relationships revealed in the social network mapping were categorized as tertiary associational ties. This associational form is characterized by fixed goals and variable membership, and the barriers to entry are based solely on the needs and desires of the organization. While this type of associational form can create social capital, it is often only used in a very narrow manner, directed to achieve the goals of the group.

The idea that creative social entrepreneurs will have larger, far-flung networks can be tested by seeing how close individuals actually are, which is measured by looking at the closeness centrality, which calculates the distance between nodes (how many steps, on average, it takes for one individual to reach all others within a network). Two findings emerged. First, there was a positive correlation between the closeness measure of individuals within the network and professional creativity in both the tertiary associations (32% at 95% degree of significance) and casual acquaintance networks (28% at 90% degree of significance). This suggests that professional creatives in Saskatoon have large, far-flung acquaintance and tertiary association networks. Secondly, there was a negative correlation between informal creativity and both tertiary associations (–32% at 95% degree of significance) and community involvement networks (–38% at 97.5% degree of significance). From this negative correlation it is possible to imply that informal creatives, in Saskatoon, operate in a small, tight community, mostly through tertiary association networks.

The second subhypothesis tested was that creative social entrepreneurs are spanners or brokers, in that they bridge or connect a myriad of individual nodes and subgroups within the larger network. The betweenness centrality measure (which measures gatekeepers who connect groups of nodes rather than simply individual nodes) was investigated for any correlation to the three types of association. The main finding was a positive relationship between the broker measures and professional creativity in both of the secondary associations (25% at 90% degree of significance for community involvement networks and 28% at 90% degree of significance for casual acquaintance networks) and primary associations (55% at 99.9% degree of significance). This relationship is the strongest that emerged in this study from the three centrality tests. One item of particular importance is that both secondary associational forms (community involvement and casual acquaintance) were strongly positively correlated with betweenness centrality and professional creativity. As secondary associational forms are the creators and incubators of social capital in a city region, this connection suggests that there is a tie between individuals with professionally focused human capital and social capital creation. More specifically, what this suggests is that, in the Saskatoon city region, professional creatives are brokers or bridgers in networks.

The effect of an individual's structural position of power in a heterarchical network and creative social entrepreneurship can be tested through examining the eigenvector centrality measure, which tests the power each node has within the networks by exploring the number of connections that a node has to other nodes, weighted by their respective number of connections. This analysis revealed two key points. First, there was a positive relationship between an individual's eigenvector centrality and their professional creativity in the context of membership in both tertiary associations (34% at 95% degree of significance) and primary associations (38% at 97.5% degree of significance). Secondly, there was a negative relationship between informal creativity and community involvement (31% at 95% degree of significance). This suggests that, in Saskatoon, professional creatives exercise structural power while informal creatives do not.

Respondents to the Creative Social Entrepreneur questionnaire were also asked to identify any concrete outcomes for the city of Saskatoon that resulted from their networks. The responses received to this question varied. In total, 19 of the 30 participants indicated that identified social networks had a strong positive impact on the creation of social capital and the collaborative governance of the Saskatoon city region. Examples of the positive impact of these social networks ranged from promoting change in people's attitudes and raising the overall quality of life (societal social capital) to the construction of two soccer centres and raising significant capital (millions of dollars) for non-profit organizations (associational social capital), and to gaining skills to benefit the neighbourhood (individual social capital). While the exact impact that creative social individuals or creative leaders have on the creation of social capital and the collaborative governance of the Saskatoon city region is difficult to measure, the survey suggested that the impact is significant.

The results from Webb's Creative Social Entrepreneurs analysis do not unambiguously support the hypothesis that creative social entrepreneurs are the primary creators of social capital in the Saskatoon city region. However, a number of conclusions can be drawn from the results. First, there is a measurable positive relationship between professional creativity (based on work, education and mobility) and all three types of social capital. Secondly, there is a negative relationship between informal creativity (based on hobbies) and the hypotheses – which suggests, at least in Saskatoon, that there are no obvious Putnamesque 'bowling leagues' to facilitate linking social capital. Entrepreneurs, in contrast, are evident but are not contributing conclusively to social capital, which may simply reflect the fact that the historically thin profit margins in Saskatoon have kept entrepreneurs strongly focused on their commercial ventures and not engaged in more diffuse efforts.

FIRM STRATEGY AND CREATIVITY

So far, this chapter has explored the individual in the cluster or the community, deconstructing the iterative relationship between the creative individual and the community in which he or she lives. But what impacts do creatives have on firms, that oh-so-important conduit of creativity and economic growth in the cluster?

What Karwandy (2008) and others call the territorial perspective on economic clustering concerns itself with the non-economic relationships and structures that can be fostered by individuals acting in close proximity. These non-market relationships lead to different behaviours on the part of community participants, not only individuals but also institutions, organizations and firms. These behaviours revolve around patterns of interaction that can create trust, common values and reciprocal openness. Theoretically speaking, these behaviours all support creative approaches to information exchange, cooperation or collaboration. The knowledge-based view of the firm covets these extra-economic benefits for their ability to assist in finding a balance between creating knowledge, integrating it into marketable items and exploiting the value inherent in these offerings. As such, Karwandy proposed that where firms engage in knowledge-based activities one might also observe elements of the territorial perspective on clustering.

Knowledge management – involving creation, integration and exploitation – is a relatively new and sometimes intangible concept that is difficult to measure directly. As a result, the research and development expenditure of each respondent is usually taken as a proxy measure of research intensity and evidence of firms adopting a knowledge-based view of management. In examining the theoretical basis for linking industrial clustering to the strategic management of firms, Karwandy (2008) modelled and tested an existing set of data (sourced through the ISRN project) that included firms operating in the Saskatoon and Vancouver bioscience and new media clusters. The relationship between the cluster and creative firms was modelled through five variables representing the territorial perspective on clustering (interaction, exchange, benefits, networking and association) and one variable representing the knowledge-based view of management (research expenditures). The proxy measures of the territorial perspective focused on the presence of trust or trust-based relationships because of their central role in defining non-market relationships. All variables, of course, are implicitly and explicitly tied to individuals – their power, how they are networked and their level of mobility and integration in the economic cluster. As the knowledge-based view is commonly associated with research activities, the variable chosen to represent it was designed to measure the research intensity of each respondent firm. Each respondent surveyed under the research protocol was asked to report their R&D expenditure by checking off a box associated with some level of expenditure. The responses were collapsed into three categories (CAN$1 million or less, more than CAN$1 million up to CAN$10 million, and over CAN$10 million) to ensure that the data were not too fragmented. The category split was based on the median of the sample.

Each territorial variable offers a different dimension of how individuals engage in creative activities through firms. First, higher levels of interaction with other agents (in this case research agencies and technology transfer centres because of their importance to the knowledge creation process) were deemed to act as indicators of a greater level of trust (the data were coded into four discrete categories: never, rarely, regularly and frequently). Secondly, the respondents' propensity to collaborate captured the range of exchanges they had engaged in. Respondents were presented with a list of nine types of exchange and asked to select all that apply – this variable does not report the volume of collaboration but rather focuses on the variety in collaborative methods. This is designed to measure the freedom with which a firm is able to engage in risky, research intensive activities. Intuitively, trust and other social constructs associated with the territorial perspective of clustering tend to facilitate the greater level of risk tolerance shown by firms that collaborate widely and creatively. The variable was coded to form an ordinal scale suitable for non-parametric analysis. Thirdly, respondents were asked what benefits they derived from their knowledge exchange relationships, on the assumption that, for these purposes, the variety of benefits was more important to capture than the volume and value of the benefits. A greater variety of benefits derived from collaboration would be expected to contribute to more successful knowledge-based firms. This variable was coded into an ordinal scale to permit the use of non-parametric testing. Fourthly, firms were asked whether they considered themselves part of a local network, on the assumption that the territorial perspective assumes that economic development is a collective process that requires willing participation. One would naturally expect firms relying on people-based strengths to acknowledge their participation in local affairs. This variable was a dichotomous variable with response categories of 'yes' or 'no'. Fifthly, firms were asked to identify whether they were members of a local association, on the understanding that participation with local associations reflects a level of connectedness and commitment to a common development objective that one associates with the territorial perspective. This variable was also a dichotomous variable with response categories of 'yes' and 'no'.

Theory suggests that firms pursing knowledge-based management strategies would be positively linked through each of these five types of engagement in order to draw upon the unique skills and aptitudes of local people. In short, the territorial perspective and the knowledge-based view of management both emphasize the potential for positive-sum wealth creation and a limitless future where knowledge is the driving force behind success.

This proposition was first tested by looking at a small data set of biotechnology firms in Saskatoon and Vancouver (n = 49). A Spearman rank correlation coefficient was calculated for the levels of interaction, the cooperation and benefits, and the firm's relative engagement in research. None of the variables was significantly related to the management variable, even when significance levels were relaxed to the 90% confidence interval. Nevertheless, the presence of benefits was positively and significantly correlated with both interactions and exchanges, which supports the notion that these variables were detecting some untraded interdependencies, albeit without any explicit relationship to knowledge-based management strategies.

The two dichotomous variables, engaging in networks and membership in associations, were assessed with the Mann–Whitney U test to assess whether the mean ranks for the two groups differed significantly from each other. The short answer was that neither was statistically significantly different, which conforms to the result found by Procyshyn (2004), presented in Chapter 6. This suggests that there is no systemic difference between firms that are networked or members of associations and those that are not. In short, engagement of this type may be a necessary condition for innovation but is certainly not a sufficient condition.

Because the non-significant results could have been due to the small sample size, Karwandy added a comparable set of respondents from the new media sector in Vancouver to bring the sample population to 92. The same test statistics were calculated, and somewhat more robust results were realized. Firms engaged in a knowledge-based management strategy (as reflected in their research investments) were positively (significant at the 95 and 90% levels, respectively) more likely to collaborate in a range of people exchanges and to identify a wider range of benefits from local engagement. They still did not show any particular greater intensity of engagement or interaction. The larger sample did however show that the intensity of interactions, the variety of exchanges used and the range of benefits were all positively and significantly correlated with each other, but not the management variable. In short, there were untraded interdependencies being generated, but the relationship with firms focused on knowledge-based strategies was not conclusive. As before, the Mann–Whitney U tests of the difference in mean ranks of those firms engaged in networks and associations were not significantly different from those not accessing these opportunities.

Karwandy's test of the proposition that the territorial or special-people-based view of economic growth is related to the knowledge-based view of management has shown that even if this holds, it is probably a more subtle relationship than could be discovered in Saskatoon. Each theory emphasizes positive-sum growth and a central role for knowledge assets. However, Karwandy's analysis generally suggests that this may not be realized by firms in practice.

CONCLUSION

While we can be confident that people matter, just how much is still under dispute. People are undoubtedly creative, in a Floridian sense and beyond. Individual effort (such as in the context of research stars) is both measurable and appears to offer real signals and knowledge transmission pathways that attract investors, entrepreneurs and firms of all sizes and shapes. Moreover, formal and informal social and professional activities connect people and can animate new ideas. It is evident that people are mobile conduits of knowledge that, when properly leveraged, can add value to a firm, organization or an economic cluster.

Our cumulative observations on the Saskatoon agri-food cluster from the people perspective, obtained through a series of studies using the diverse methodological approaches outlined in this chapter, do point to the need for better modelling and measurement instruments in the field of people-based cluster research. Identifying and analysing creative cluster leaders and understanding how they influence growth and change in a given cluster is important – while Webb (2009) offered some tantalizing insights, they are far from definitive. Another key component in all of this is the Karwandy (2008) study that initiates a deductive process by which subsequent research can focus on the causal pathways that translate the efforts of creative people into commercial success. Finding these key pathways between people, activities, organizations and firms that link performance to socio-economic outcomes and then developing methods for more effectively measuring cluster success will be vital to exploit better the power of people-based strategies in complex and fluid networks. While much of the literature suggests that this should only be a concern for the high-profile creative industries, the work discussed in the last six chapters at a minimum shows that the problems and opportunities of knowledge-based growth extend well into the traditional economic sectors, in this case the global agri-food sector.

9 Moving Innovation Studies from Comparative Statics to Networked Dynamics

Interactions between humans, even those taking place in economic contexts, do not operate in a vacuum. They are inherently social in nature, involving formal and informal chats, exchanges of information and collaborative work with colleagues, and are frequently in tightly connected communities, working and living as next-door neighbours. These recurring patterns of exchange and interaction bind people together into social networks. How people operate or act in these networks is highly dependent upon their values, interests and beliefs and the assets (physical, social and informational) that may be available to them at any time.

Standard economic theory traditionally ignores the nuanced nature of these arrangements and the underlying factors that make them fundamentally interdependent and dynamic. Most theory typically treats these factors as mutually exclusive. To be fair, the individualist, static framing of most theory is not simply because people have not understood the potential for dynamic effects. In the first instance, the mathematical logic used to drive the hypothetical–deductive scientific method primarily used in microeconomics was for the most part not up to highly interactive conceptual modelling. Perhaps as importantly, those interested in applying the models with data were constrained because the computational tools were especially weak. Both of these barriers have at least partly been surmounted, opening up new approaches to such complex systems.

The theory began to evolve first. In the late 1980s, a group of about 20 people came together at the Santa Fe Institute to talk about the economy as an evolving, complex system. The group included economists, physicists, biologists and computer scientists – all invited by two theoretical mavericks,

Kenneth Arrow and Philip W. Anderson. This multidisciplinary group was gathered to stimulate new ideas under the rubric of the 'sciences of complexity' and to discuss new ways of thinking about economic problems (Arthur *et al.*, 1997). The efforts of the Santa Fe Institute and its affiliates focused on the notion of complexity in the economy. Traditional economic theories do not account fully for the dynamic and complex interaction of people, information and other factors in an economic setting. This group undertook some of the early foundational work of translating the emerging concepts of complexity, based on feedback loops that lead to a range of effects one could not anticipate or forecast based simply on the examination of each of its component parts. These systems effects – often called 'emergent properties' – including tipping points, power–law relationships and long-tail distributions (sometimes called 'black swans'). In short, the theory demonstrates that, in many cases, the system is greater than the sum of the parts, which fundamentally challenges the reductionist approach commonly used to study such effects.

At about the same time, number and graph theory was advancing to develop new ways of handling and undertaking complex calculations of interrelated data. While we have learned a lot from the application of traditional statistics and econometrics, the work-horse regressions that have delivered such insights are based on a fundamental assumption that the independent variables are not related to each other – when they are, our estimates fail to deliver unbiased results. In a world of relational data, where data are structurally interconnected, new methods of analysis are needed. Number and graph theory, as translated into the tools of social network analysis, or more commonly called SNA, have opened up whole new areas of investigation. While we have yet to reach an end point on this path, already there are a large number of share-ware, open-source software programs that facilitate organizing, visualizing, statistically describing, validating and ultimately simulating the effects of highly interactive complex systems.

These new theories and tools have only recently been adapted and adopted in the study of clusters or regional innovation systems. While the study of localized economies was initially undertaken by economic geographers, business-school management researchers and some sociologists, the field is now flooded with a much wider array of disciplines and methodologies, many of whom are attempting to apply complexity theory and the related analytical tools to the study of innovation (Ryan, 2004). Most analysts now agree that economic agglomerations are not simply driven by competitive firms as theorized by Porter (1990). Rather 'clusters are dynamic and complex entities ... complex adaptive systems; not self-sustaining but rather self-reinforcing as proposed by economic complexity theory' (Ryan, 2004, p. 3). Key factors such as knowledge – especially tacit, co-produced, contextually rich knowledge, which only has meaning in communal settings – and social capital, which by its nature implies interconnectivity – are not viewed as vital underpinnings to innovation. Given that social context or the social system is now part of facilitating the innovation process, it is necessary to bring in new theories and tools to understand their role. Social dynamics or systems dynamics (often interchanged or considered to be closely related in theoretical discourses) are fundamentally concerned with changes over time (as no social phenomena ever truly happen instantaneously). The key feature, as illustrated in Kline and Rosenberg's (1986) chain-link model of innovation, is the role of feedbacks within the given system.

THE THEORY AND PRACTICE OF DYNAMIC SOCIAL SYSTEMS

The key component underlying innovation is knowledge. Knowledge itself is differentiated in terms of both its structure and its transferability (Malecki, 1997; Ryan and Phillips, 2003; Ryan, 2008). The ability to identify, access and leverage knowledge in cluster settings suggests that actors need to be proactive and flexible in their strategies in order to work with a multitude of organizations and firms. Nonaka and Takeuchi (1995) described the knowledge imperative as a dynamic, social and continuous process involving the acquisition, accumulation, creation and exploitation of new knowledge. At the heart of what these authors suggested is a dynamic process that involves organizational learning, which is a highly social phenomenon (Carley, 1986). The knowledge system – involving people, institutions, organizations, ideas, rules and exchanges – is embedded within a dynamic economic context. This inevitably leads to a complex system comprised of interconnected social subsystems, some which might be decomposable or nearly decomposable but many that will not be (Simon, 1955).

The dynamic and ever-changing nature of social systems suggests that traditional, static input–output models for evaluation or quantification will be limited at best. When we take into consideration the complexity of any given economic or social system, traditional methodologies fall short in terms of measuring such phenomena. In fact, most are unable to incorporate much in the way of knowledge factors, collaborative models and structures, intermediate performance measures, proximity, differing levels and types of social capital, and the evolution of that social capital over time.

Social networks represent a way to more directly and effectively examine innovation in clusters. According to Castells (2000, pp. 501–502), a network-based social system is a 'highly dynamic, open system, susceptible to innovating'. Castells (2000, p. 422) further suggests that 'social networks, of different kinds, powerfully contribute to ... innovation, and to its dynamism, ensuring the communication of ideas, the circulation of labour and the cross-fertilization of technological innovation and business entrepreneurialism'.

Present day SNA has grown out of a diverse range of theoretical traditions. Its underpinnings can be traced to three roots, in graph and number theory, group dynamics and inter-personal relationships from sociology and anthropology. Over the past several decades, SNA – or some form of it – was used earliest and most often in epidemiological studies (to map the evolution of disease or infection) and in computational science (to examine people-to-technology linkages). More recently, the tool has been applied to management studies for exploring the interaction among economic agents and in evaluating organizational governance.

SNA as a research tool can illuminate the dynamic interactions between and among knowledge, people and institutions and can essentially make previously invisible linkages visible (Mead, 2001). As economic growth is highly dependent on linking to and manipulating the global flows of knowledge, SNA can identify the spatial coordinates of the institutions that possess knowledge stocks and determine the direction of the flows of knowledge. Ryan (2008) suggested that SNA can be utilized to deconstruct the institutional activities that are responsible for knowledge development. SNA is distinctive in that it is most interested in 'the relationship' that exists between a set of actors or agents. Relational ties can be characterized as kin-based or role-based, or may represent affective relationships (who likes whom). In graphical analysis, agents or actors (sometimes called nodes) within a network are connected by edges or ties (which can be undirected, binary or reciprocal) or arcs (which are defined as directional and not reciprocated). The capacity to incorporate weighted data to characterize the quality of a relationship is possible with some software. Relationships between agents can include discrete ties between actors but also may include an individual's ties to other actors, nodes or events. In terms of the data set, the latter is commonly referred to as two-mode data.

As a tool for analysis, SNA views actors and actions as interdependent units and posits that relational ties between agents provide channels for transfer or flow of resources that can create opportunities for or constraints on individual action (Wasserman and Faust, 1994). SNA can help to identify boundary spanners, gatekeepers and knowledge bottlenecks, and as well as under- and over-utilized individuals or organizations in a system. The relationships (however they are defined) among actors or agents make up what is known as the network structure. This structure may vary from being quite dense with many connections among agents (relative to the total number of possible links) within the network, to being sparsely knit with few links connecting agents. From a structural standpoint, it is just as important to identify gaps within a network as it is to identify and quantitatively assess links. Burt (1992, 2000) posited a Structural Hole Theory, which explored the nature of network gaps. He suggested that such gaps may, in fact, be a positive network attribute. Structural Hole Theory predicts a negative association between networks that are dense (without structural holes) and performance.

Burt's theory reflects aspects of Granovetter's (1983) concept of 'the strength of weak ties'.

The quantitative measures that can be derived from network studies are a very attractive component of SNA. The SNA process, which is readily facilitated through a variety of SNA programs and tools (e.g. UCINET,[2] INFLOW[3] and ORA[4]), both visualizes networks or systems and their inherent links and applies quantitative valuation to the network, its nodes (actors or agents) and the links. These quantitative values are very useful in comparing and contrasting systems and in evaluating the evolution of a network over time.

The most commonly used network-level measure in SNA is density. Density is used to quantitatively analyse the structure of the broader network or system of interest. The measure gives insight into the nature of the connections among actors, affiliates, their location and other qualitative factors. According to Wasserman and Faust (1994, p. 181), density quantifies the 'knittedness' of a network. In mathematical terms, the density of a network is the proportion calculated as the number of all ties occurring in a matrix divided by the number of all possible ties (Knoke and Kuklinski, 1982, p. 45). The measure takes on values between 0 (an empty network with no connections) and 1 (a network that is completely connected). A higher-density measure is often correlated with more optimal network function or structure. This, however, is not always the case. Valente *et al.* (2007, p. 15) explored density as it relates to community coalitions in health-programme delivery, finding that too much density may indicate network-centric connections that 'do not provide sufficient pathways for information and behaviours to come from outside the group'. In contrast, a low density may make a network less effective at mobilizing resources for adoption of prevention strategies. At both extremes, path dependency or group-think can limit network-based learning.

Beyond density, there is a wide range of network- and actor-specific measures that help to define the emergent properties of networks. One commonly used measure in SNA is centrality. The concept of centrality refers to the importance of a particular actor in the context of an entire network. In general, centrality measures are used to describe and measure properties of actor location (Wasserman and Faust, 1994). Centrality, applied at the node

---

[2] UCINET (current version 6.0) is a comprehensive package for the analysis of social network data as well as other mode 1 and mode 2 data. Integrated with UCINET, the NETDRAW program can provide diagrams of social networks. In addition, the program can export data to MAGE and PAJEK for visualization of graphs (Borgatti *et al.*, 2002).

[3] INFLOW performs network analysis and network visualization in one integrated product (Krebs, 2005).

[4] Organizational Risk Analyzer (ORA) is a network analysis tool that detects risks or vulnerabilities of an organization's design structure. ORA utilizes over 50 measures, categorized by which type of risk they can detect (Carley and Reminga, 2004).

level, involves a family of measures (currently 70 and growing), each of which addresses a specific theoretical proposition. Essentially, centrality measures the relative importance of an individual actor based on their location within the system. It is an index based on the number of linkages divided by the number of total linkages to other network actors. Four main centrality measures are most commonly used: (i) total degree centrality; (ii) betweenness centrality; (iii) closeness centrality; and (iv) eigenvector centrality.

Total degree centrality refers to the informal leadership capacity of a network actor or agent according to the number of ties that the node has. Total degree centrality measures the degree to which one individual or actor is connected to other network actors. According to Mote (2005), a higher total degree centrality implies a higher level of network activity. The measure is dependent on the size of the network; thus, the total degree centrality measure of a given node or actor is normalized by network size (Wasserman and Faust, 1994). This measure is 0 for any actor that has no connections with other network actors. The total degree centrality measure is 1 if an actor is linked with every possible partner.

Closeness centrality builds on degree centrality by looking to the nature of the distance between nodes. Wasserman and Faust (1994, p. 184) referred to this measure as a 'distance function' where the measure reflects 'how close an actor is to the other actors in the network'. Mathematically speaking, closeness centrality calculates how many steps (on average) it takes one actor to reach all other nodes or actors within a given network. Like the total degree centrality measure, closeness centrality measures range between 0 and 1 and can be viewed as inverse average distances between one actor and other actors or nodes within the network (Wasserman and Faust, 1994).

Interactions or connections between nodes or actors within a network may depend on other actors or sets of actors that exist on the path between them. These 'other actors' may have some control over these connections or interactions (Wasserman and Faust, 1994). Betweenness centrality is a measure of this 'influence' and, based on a normalized measure that ranges between 0 and 1, identifies those actors who occupy the role as a gatekeeper or intermediary within a given network. Valente (1995) posited that betweenness centrality quantifies how often an individual actor is located on the shortest path between other actors and subnetworks. A higher betweenness centrality measure implies a greater level of control over information pathways of interaction between actors or nodes.

Finally, the eigenvector measure quantifies power. It calculates an actor's centrality relative to the sum of the degrees of the actors or agents to which they are connected (Bonacich, 1972; Carley and Reminga, 2004). The actor or node with high eigenvector centrality is connected to many actors who are themselves connected to many actors, thus multiplying their risk and opportunity within the network – a high eigenvector rating implies relative power in a network. The measure is derived not from the quantity of connections but rather from the relative importance of an actor's connections. Thus, the quality of connections is key in terms of this measure. In many cases, an actor having a higher eigenvector ranking suggests that this individual or organization may have greater diversity in terms of sources of and uses for information.

Ryan (2008) parsed out these measures, extrapolating from theory and applying the terms in the context of evaluating performance of research networks. Table 9.1 refers to the four key centrality measures and how they can be used in evaluating networks or systems.

SNA is increasingly being used to map innovation networks. SNA enables a researcher to identify the relative position, function and power ranking of the individual actors, nodes and subnetworks in a quantifiable and graphical manner. By understanding how networks are shaped and the role that specific institutions (including government) or individuals play within those networks, policy makers are able to make informed decisions on how the social capital accrued within the network can be leveraged to foster innovation. In short, this tool is useful in examining the dynamics of socio-economic systems.

DYNAMIC SYSTEMS ANALYSIS IN SASKATOON

Over the course of a decade, the Saskatoon-based research team conducted a series of analytical studies on the Saskatoon agri-food system and the broader community. Quantitatively, this work represents some unique steps to better qualify the nature of the community, its structure and its capacity to grow and evolve as an economic system (Procyshyn, 2004; Karwandy, 2008; Ryan, 2008; Webb, 2009).

The Saskatoon-based work drew on a number of previous studies done elsewhere to inform the methodological approach. One particularly important piece was by Theodorakoupoulou and Kalaitzandonakes (1999), which developed an approach that incorporates both knowledge and social capital into a methodology of examining networks in the plant biotechnology sector. As discussed in Chapter 6, Theodorakoupoulou and Kalaitzandonakes mapped the national knowledge networks of the industry in the USA and EU by analysing, comparing and investigating the impacts of the network structures, represented by public–private research agreements. The goal of the study was to measure the effect of the network position of each participant on innovation performance of the industry. Information was gathered on public–private research agreements in both the USA and EU, using CRADAs in the USA and the Framework Programmes grants in the EU. Both sets of knowledge networks (USA and EU) were analysed separately using network analysis. Particular attention was paid to the density of each network and the measure of betweenness centrality. The authors used these variables as inputs

Table 9.1. Typology of centrality measures. (Adapted from Ryan, 2008.)

Centrality measure	Definition	Network-level application (centralization)	Node-level application
Total degree	An actor or principal with higher total degree is identified as a hub or connector within the network	To determine the overall intra-network connectedness and determine to what degree the network revolves around one actor or principal	To identify central actors or principals (those that are more highly connected) in terms of position and activity; those actors or principals that are 'in the know'
Betweenness	An actor or principal with high betweenness is identified as a broker or bridge and can connect or disconnect groups within the network	To determine the overall knowledge flow capacity within the network and to what degree the network revolves around an actor or principal that connects network subgroups	To identify those actors or principals with the capacity to facilitate or break knowledge flows within the network; those that link subgroups within the system have potential influence
Closeness	An actor or principal with higher closeness measure has less total distance to all other nodes	To determine how long it will take a message or information to spread within a network or system	To identify those central actors or principals that have the shortest distances to others within the network; those actors or principals with optimal transference power
Eigenvector	An actor or principal with a higher eigenvector has multiple connections with others with multiple connections	To determine the overall level of cohesiveness of the network and to what degree the network revolves around actors or principals that are connected to other central actors	To identify those central actors or principals that have more connections to other highly central actors or principals; those actors or principals with influential power

to a regression analysis, which demonstrated that density was inversely related to successful commercialization and that network structure and position of actors and institutions had a significant role in determining the power of the actors involved and their ultimate commercial success.

Using this earlier work as a guidepost, a number of Saskatoon-focused studies were undertaken. Table 9.2 provides an outline of approaches used to evaluate the dynamics of economic exchange systems, including clusters.

Table 9.2. Typology of innovation-based performance evaluation approaches.

Author(s)	Focus	Measures[a]				Inputs	Output measures	Parameters			Dynamic or static
		Density	TDC	BC	EC			Know-ledge	Social capital	Other	
Theodorakou-poulou and Kalaitzand-onakes (1999)	USA and EU	✓	✓			Number of joint projects/ collaborations	Degree of activity and network positions	✓	✓	X	Static over several years
Procyshyn (2004), Procyshyn *et al.* (2005)	Cluster	✓	✓			Survey factors/ cluster-based activities (services, research, funding, etc.)	Innovative index (firm), density and centrality (region/cluster level by activity)	✓	✓	✓	Static
Coenen *et al.* (2006)	Regional cluster	✓	✓			Publications and patents	Centrality and density (publications and patents)	✓	✓	✓	Dynamic comparative across regions over intervals of years

Ryan (2008)	Regional research networks	✓	✓	✓	✓	Publications, patents, other connections	Centrality, density, awards, other connections	✓	✓	✓	Dynamic comparative across regionally based networks over time
Webb (2009)	Creative social entre-preneurs	✓	✓	✓	✓	Survey factors/ qualitative data	Centrality, density, awards, other connections	✓	✓	✓	Static
Phillips *et al*. (2009), Boland *et al.* (2010)	Global plant genetic resources	✓	✓	✓	✓	Secondary data collection	Centrality, density, sensitivity analysis	✓	✓	✓	Dynamic, static and sensitivity analysis; comparative across jurisdictions
Ryan (2011)	CLS Network of Scholars	✓	✓	✓	✓	Publications/ survey factors	Centrality, density, awards, other connections	✓	✓	✓	Dynamic comparative over time

[a]TDC, total degree centrality; BC, betweenness centrality; EC, Eigenvector centrality.

Procyshyn (2004) explored the Saskatoon agricultural biotechnology cluster using new (at the time) metrics to assess cluster-based functions and to assess the community's capacity to become a broader life-science cluster. The study method leveraged survey data drawn on interviews with representatives from the cluster's core actors. The purpose was to determine connections among actors within the industry in the area and to evaluate the structure of the Saskatoon cluster and actor-based activities according to five cluster-based functions: (i) research and development; (ii) fee for services; (iii) financing; (iv) exchange of high-quality personnel; and (v) networking. Using the SNA tool, the density and centrality of the network of actors were measured. When the data were disaggregated by cluster-based function, Procyshyn discovered that networking had the highest density of all five functions, suggesting that, at the time, the local cluster was still in the innovation stage of industrial development (Lundvall, 1992). Procyshyn then examined Saskatoon's central actors to determine whether they provided differential functions to innovative firms (as opposed to non-innovative firms). Only three central actors were linked (with statistical significance) with activities supporting highly innovative firms: NRC-IRAP was identified as a source for the provision of research and development and related support; Ag-West Biotech was identified as a source for financial exchanges; and NRC-PBI appeared to offer differential networking services.

In order to understand networked innovation in the Saskatoon region relative to the broader global context, Ryan and Phillips collaborated with colleagues at the Centre for Innovation, Research and Competence in the Learning Economy (CIRCLE) at Lund University in Sweden to contrast and compare socio-spatial patterns of knowledge creation in a biopharmaceutical region (Scanian) with the Saskatoon agri-food biotechnology clusters (Coenen *et al.*, 2006). An important distinction between these two regions is that the Scanian biopharmaceutical region is distinguished as a conduit of upstream, analytically based knowledge, while the Saskatoon agri-food region knowledge sphere is more 'synthetic' or downstream in nature. Various dissimilarities were noted and were not unexpected, based on differences in historical technical regimes and overall sectoral innovation systems dynamics of the respective regions. One significant distinction of note is that, when compared with the Scanian region, Saskatoon's agri-food biotech cluster is less spatially distributed in terms of co-patent collaboration patterns, even though co-authorship patterns across both regions were quite similar. Drawing on Florida's notion of the creative class, Webb (2009) approached the analysis of the Saskatoon cluster from another direction. He investigated the hypothesis that creative social entrepreneurs facilitate collaborative governance in the Saskatoon city region by being the primary creators of social capital. Webb sought to test whether governance, both its quality and form, is a key factor in facilitating social, cultural and economic development at community level. As discussed in Chapter 8, his study leveraged survey data from 30 creative social entrepreneurs representing government, business, civil society and the university in Saskatoon. Webb's survey was designed to measure the level of creativity (professional and informal creativity) and entrepreneurial capacity of the interviewees, as well as to map their social networks. Using three measures of centrality (closeness centrality, betweenness centrality and eigenvector centrality), the author examined the impact that creative social entrepreneurs play in social capital creation in the cluster. The results from this analysis did not explicitly support Webb's hypothesis that creative social entrepreneurs play a key role in social capital creation and the facilitation of governance at the cluster level. According to Webb (2009), not all innovators impact on social capital creation. However, those professional innovators that do act socially are positioned in community networks to be the primary creators of social capital.

The other Saskatoon-based studies that used SNA tools are discussed in more detail below.

CASE STUDIES OF STRUCTURE AND FUNCTION IN THE SASKATOON COMMUNITY

Ryan (2008) undertook a major study of the knowledge exchange structures in the Saskatoon-based agricultural biotechnology cluster, seeking to determine whether they have converged on a potentially cost-effective and efficacious structure. The array of collaborative ventures, joint initiatives, licensing agreements and subsequent patents that have been assembled or developed between actors – and between star scientists – over the past few decades has enabled the community to deliver a number of world and market firsts. Two recent developments were investigated.

First, the community structure identified and examined by Phillips and Khachatourians (2001) and discussed in Chapters 4, 6 and 8 is now being influenced by a broader range of new actors with new interests. There has been significant reinvestment in the Saskatoon life-science sector, for example, through the Genome Canada (GC) initiative. As of 2011, the Federal Government has invested almost CAN$900 million into GC for the purpose of developing genome centres across Canada. Saskatoon's agricultural biotechnology community welcomed these investments, embracing the opportunity of establishing an agricultural genomics centre within the region. The initiative has come to involve government, university and a range of industrial partners. GC has made a range of investments in Saskatoon in animal and plant genomic research projects valued at more than CAN$80 million over the decade.

The Project on Functional Genomics of Abiotic Stress in Wheat and Canola Crops (CAN$419.5 million, hereafter called the Abiotic Stress

Project), an initiative of GC managed by Genome Prairie in Saskatoon, is a good illustration of the new pressures that are emerging. This project involved 21 principal investigators across five universities, six provinces, three federal laboratories and three international partners seeking to investigate cold tolerance of canola and wheat varieties. In addition to requiring more scientific collaboration, this and the other GC genomics projects are directed to link up with scholars in Genomics, Ethics, Environment, Economics, Law and Society (GE^3LS) studies. GC states that these new relationships are designed to ensure that the scientific research is complemented and made more socially appropriate through interaction with the examination of associated contextual issues.

While major collaborative research programmes offer substantial potential benefits in terms of research outcomes, they create potential for higher implicit and explicit transaction costs. Government policy, as reflected through the granting councils – Canadian Institutes of Health Research (CIHR), CFI, the Natural Science and Engineering Research Council (NSERC) and GC – now as a matter of form require significant collaborative relationships in most new major research projects. The project initiation and development process (as for the genomics projects) has become significantly more resource intensive, involving multiple actors across multiple disciplines in often different geographical regions. There are various stages to the project development process: search, negotiation (re-negotiation) and the ongoing developmental processes associated with sustaining the project. Substantial transactions costs (both implicit and explicit) often impede efforts to develop projects and to commercialize outcomes.

In order to mitigate costs, GC project activity in agricultural genomics appears to have clustered in and around the Saskatoon region (Fig. 9.1). Phillips (2002) suggested that one reason for the clustering in Saskatoon might be to exploit local social capital – in terms of network connectivity – to limit transaction costs. One illustration of this is that some projects appear to have been built upon or are extensions of existing relationships among principal investigators.

To test the theory, we constructed and examined the network structures of three key projects – the Abiotic Stress Project, VIDO's Animal Mucosal Vaccine Development project (CAN$26.9 million) and NRC-PBI's Designing Oilseeds for Tomorrow's Market project (CAN$17.5 million, hereafter referred to as the Brassica Genome Project) – and the search, negotiation and ongoing costs associated with each stage of project development in order to determine whether these projects, whose actors have high levels of social capital, will have lower transaction costs, which is posited to contribute to more successful collaborations. Ryan and Phillips (2003) assessed this through the quality or number of activity-based linkages, while Procyshyn and Ryan (2003) examined the influence of historical and ongoing relationships and a high network intensity.

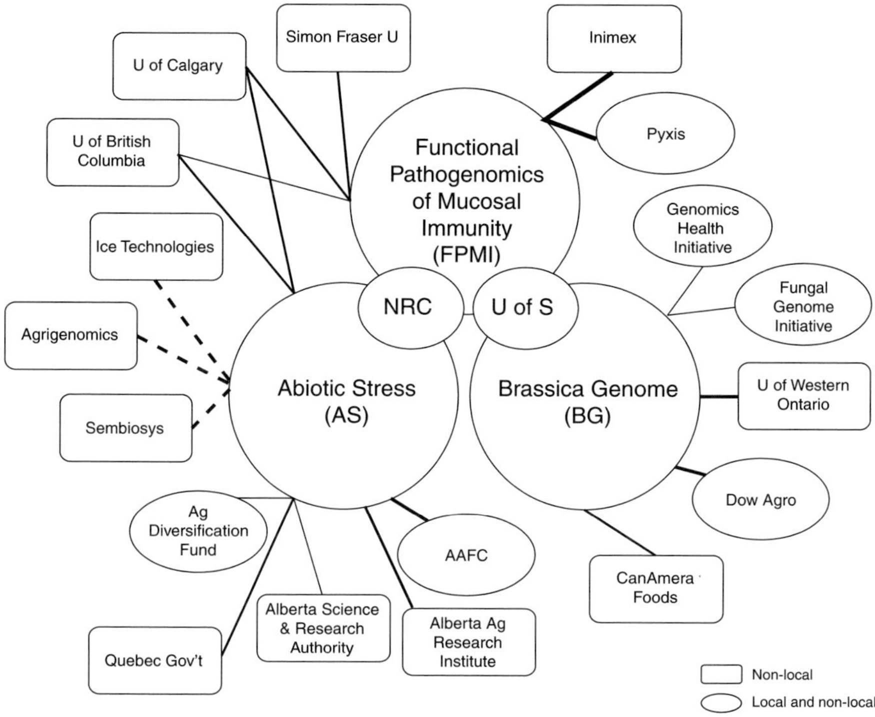

Fig. 9.1. Collaborations related to GC genomics projects in Saskatoon. U, University; U of S, University of Saskatchewan.

Figure 9.1 shows the linkages these three agricultural genomics projects had with both local and non-local organizations and firms. The weight of line indicates the nature of the relationship. A heavier line denotes more connections (i.e. funding, in-kind contributions, research links, exchanges of high-quality personnel and services provided). Secondary or indirect linkages are illustrated by dotted lines. The results serve as a preliminary step in illustrating the structural context for project development and management.

A survey instrument incorporating both qualitative and quantitative parameters was developed to determine past, current and anticipated relationships related to research and commercialization involving the genomics research in the Prairie region. The instrument was designed to shed light on the three stages in project development, focusing on search costs, negotiation costs and ongoing costs. Additional questions investigated intellectual property strategies, the costs incurred in the search, negotiation and monitoring of intellectual property rights, and commercialization plans. The survey instrument was administered between July 2003 and early 2004. The target population consisted of 16 projects, which were submitted for consideration over the first two rounds of competition. Of these, only six were funded. We conducted interviews with five of the funded projects.

For the purposes of this project, we focused data analysis on those projects that were geographically concentrated in and around the Saskatoon region with activities centred on agricultural biotechnology: Abiotic Stress (AS), Brassica Genome (BG) and the Functional Pathogenomics (FPMI) projects. These projects represented CAN$54.3 million in aggregate awards by GC during the first and second national competitions. The study offers an analysis of the quality of networks associated with each of the three projects according to the number of actors in the network, the number of linkages per actor and the overall quality of those linkages. Ryan (2008) reported that the FPMI exhibited the highest level of network quality of all projects. FPMI's relatively strong linkages were based on long-standing relationships, which appeared to contribute to reduced transaction (implicit and explicit) costs associated with the search and negotiation phases of project development. This is in sharp contrast to the AS project, which was awarded funding after two rounds of negotiation and only after two discrete proposals were combined into a larger structure. As indicated through the interview process, the actors in this case did not have a history of working together, which was judged to contribute to the higher transaction costs for the AS project. Moreover, in the case of FPMI, strong linkages among a relatively few actors appear to have enabled this project to obtain a much higher absolute and relative level of funding, while FPMI's implicit and explicit costs appear to have been significantly lower than those of AS and BG.

Overall, it appears that transaction costs are at least partly a function of the number of actors, the strength of the linkages and the history of previous relationships or collaborative ventures.

Meanwhile, the establishment of CLS in Saskatoon represents a new development that has added and continues to add complexity and nuance to the evolving network dynamics and embedded structures of the local innovation cluster. Ryan (2011) and Ryan *et al.* (2012) conducted a study of the CLS scholarly network to augment a more traditional social and economic impact assessment of the facility's productive capacity. The study applied the SNA methodology and combined it with statistical analysis to assess the research network. To add value to the analysis, Ryan framed the study in the context of the Treasury Board of Canada Secretariat's Results-based Management Lexicon and the Government of Canada's Science and Technology Framework. These frameworks identify the importance of knowledge, people and entrepreneurial advantage for national prosperity.

Using a data set of 4000+ publications generated by CLS-related researchers between 2005 and 2011, Ryan (2011) and Ryan *et al.* (2012) conducted a SNA (targeting the density and centrality measures) to evaluate the growth and evolution of the network of scholars affiliated with the institution over time (year by year). First, she discovered that connectedness among CLS scholars is unambiguously growing. Further analysis showed that there is a positive relationship between high centrality measures (scholars that are more central within the CLS network such as team leaders) and the goal of knowledge generation. The relationship between higher centrality measures and entrepreneurial-focused research are negative and not as strongly correlated as is the case with knowledge advantage, while correlations between centrality measures and production of high-quality personnel (e.g. graduate students, post-doctorates and technicians) was negative, suggesting that those occupying positions of power are either disinclined or too busy to supervise and mentor new researchers. One encouraging result was that there is a strong, positive correlation between higher total degree centrality measures (those scholars that are more connected) and use of CLS facilities – the facility is attracting people that are able and willing to collaborate. Given this, Ryan was somewhat surprised that there was no detectable relationship between central placement in the CLS world and the capacity of the individual to generate peer-reviewed grants.

While the CLS is in its early stages of development, this research revealed that both visually and statistically there is a growing community forming around the infrastructure. The knowledge being generated by the system or network is accelerating.

A CASE STUDY OF SASKATOON'S ROLE IN GLOBAL RESEARCH MANAGEMENT

As discussed in Chapter 5, the pipeline–buzz framing acknowledges that, while place and proximity matter to innovation, the activities of a given region or cluster are not cut off from activities in the rest of the world. In fact, the capacity for regional or cluster innovation often depends on the leveraging of and extraction from global sources of knowledge and resources. A common theme that underscores the recent collaborative-oriented institutional and global perspectives is the need for the utilization of public–private partnership (P3) organizations to forge the links, either globally or institutionally, between various organizations and networks to facilitate knowledge generation and diffusion (Bathelt *et al.*, 2004; Etzkowitz and Ranga, 2010).

This global phenomenon has been explored in a series of studies of the global pulse breeding and R&D system that is linked to Saskatoon. Phillips *et al.* (2009) and Boland *et al.* (2010) used SNA to undertake a comparative policy analysis on the global pulse breeding R&D network of 248 actors. A quick overview of the results showed that the P3 organizational model, while representing only 18% of all the network actors in question, represented all the actors that occupied the most critical structural places and provided the key linkages needed to sustain global pulse research system (Fig. 9.2; P3s

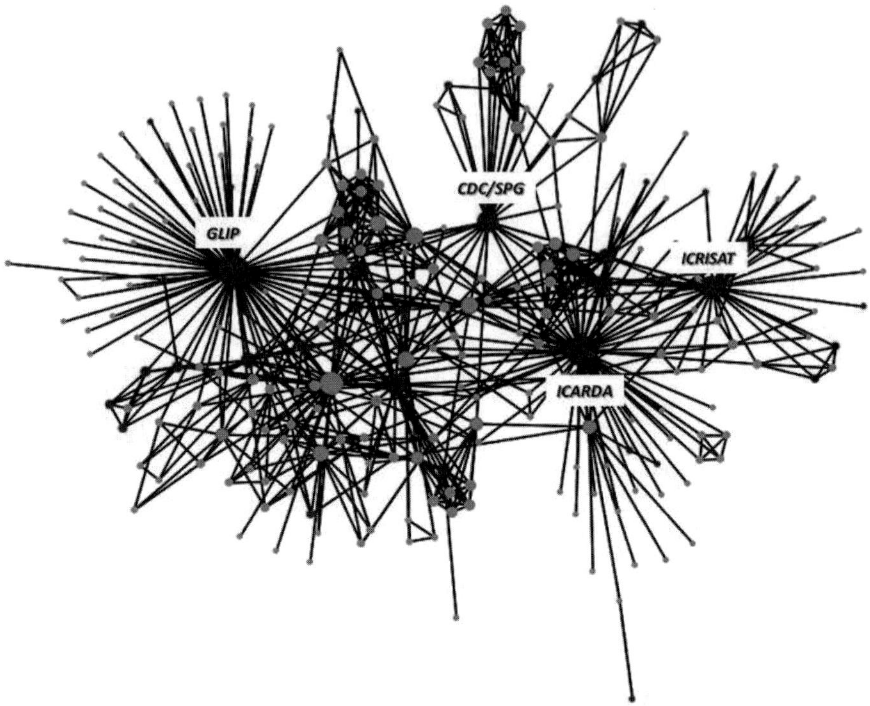

Fig. 9.2. The global pulse breeding and R&D system.

are represented by darker nodes). The four most central actors in the global system also happen to be P3s: the Grain Legumes Integrated Project (GLIP) in the EU; the International Centre for Agriculture Research in the Dry Areas (ICARDA) and International Crops Research Institute for the Semi-Arid Tropics (ICRISAT), both international organizations connected to the Consultative Group on International Agricultural Research (CGIAR); and the Crop Development Centre-Saskatchewan Pulse Growers (CDC-SPG) partnership located in Saskatoon.

Using SNA-based analysis, we tested for vulnerability in the system by removing the four highest ranked actors (according to centrality measures). This leads to a 25–98% impairment of network function (Fig. 9.3). The Canadian system actually becomes disconnected from the global system, with its subnetwork in the upper right-hand quadrant of the figure.

The analysis returned other interesting results. The P3 is the focal point for a comparative policy analysis between the Australian and Canadian national systems of pulse R&D. The two subsystems are the primary research contributors and competitors in export markets for pulse crops. Each system depends on the P3 to anchor the system nationally and to connect the national system into the regional and global networks. The Canadian system

Fig. 9.3. The global pulse network without key P3s.

relies on a single P3, the CDC-SPG, a producer-funded initiative, to anchor its R&D network, while the Australian system has developed a range of P3s that are centrally placed in three interdependent and highly specialized R&D networks. In some ways, the Canadian system represents a bottom-up and local approach to R&D, while the Australian system is the direct result of multiple national R&D policies and programmes that specifically utilize P3s as a method of distributed governance to develop and manage collaborative R&D networks. Boland *et al.* (2010) concluded that the Australian system exhibits deeper specialization and greater integration into the global system compared with Canada, which is highly centralized and dependent on a single organization to link to the global system.

Phillips *et al.* (2008b) attempted to determine the impact of these alternative approaches in Australia and Canada. They looked at the research impacts in the context of new crop varieties commercialized relative to the market size, concluding provisionally that the Australian model was translating into a more rapid and significant uptake and use of new seed technologies. More work is clearly needed to assess how different network structures affect commercial performance.

CONCLUSIONS

Many governments around the world are committed to knowledge-based growth, including for the agri-food sector. Most of these strategies start with significant upstream investment in research and development. These investments in science and technology are designed to generate something beyond science alone. To understand their impact requires more than the usual comparative static analysis that permeates the gains-to-research economic literature. As this chapter has shown, there is now a strong theoretical base and a new set of tools that can be used to extend the investigations to incorporate dynamic systems with substantive feedback loops.

Public and private investment has had a major impact on the Saskatoon region, its economic development and the generation of new knowledge. Government-funded infrastructure and projects, like the CLS and the GC projects, for instance, offer the context for scientists to conduct their work and to do things that could add socio-economic value. It is important to remember that scientists seldom work in isolation. Today, most are part of larger networks. Communities that can generate bigger net effects for the individuals and institutions that engage are likely to be more successful innovators. Understanding the structure and scale of these scientific networks provides insights into the impacts of scientific investment on the scientific community (locally and globally) and the potential opportunities that may be realized by leveraging network-based assets. Some assert that these networks generate social capital that enhances and leverages the physical and financial capital directed at scientific discovery and amplifies the momentum to introduce new products and processes to the market. The evidence is at best preliminary but offers some promise.

The complexity of factors that are inherently a part of any regionally based science effort needs to be taken into consideration in the development of methodologies for assessment. The range of studies and approaches adapted and applied to the Saskatoon cluster over the past decade, predominantly utilizing SNA, looks to offer a viable and valuable way to examine the relationship between collaborative efforts among actors or agents within a region and key social and economic outputs and outcomes, be they immediate, intermediate or long term. These approaches, which continue to be refined and developed over time, appear to be more efficacious than traditional input–output economic methods in terms of evaluating productive capacity in economic aggregations and networks.

10 Evidence-informed Policy for the 21st Century

The end of all our exploring will be to arrive where we started and know the place for the first time.

T.S. Eliot

It is evident that economic agglomerations are complex phenomena that are difficult to track and measure over time. The evolutionary paradigm that is part of economic geography offers some insight into the phenomenon of clusters, their landscapes, how they evolve and how they transform over time (Martin and Sunley, 2003; Asheim and Gertler, 2005; Boschma and Frenken, 2006; Boschma and Martin, 2007, 2010). But these efforts feel incomplete. It appears that a combination of comparative statics, melded with selected dynamic methods such as SNA, as encapsulated in the longitudinal study of the Saskatoon cluster, might help us forge an observable and manageable understanding of the economic phenomenon.

The metaphor of the cluster life cycle, analogous to the biological representation of living matter, appears to be a good place to start to conceptualize things. The life cycle metaphor captures the notion of time, evolution of markets and economic change through four or five stages. It is understood, implicitly, that there is a complex interaction of factors at work that drives innovation in a given economic setting. Scientific, technological or industrial specialization is presumed to be at the heart of it.

Social change itself is an elusive concept and is very difficult to map. It is highly contingent on the actions of ordinary people, often doing ordinary things. As a society, we sometimes embrace change but at the same time we often vehemently oppose it. Social movements, political campaigns and business strategies are often structured around a perceived need for change or in response to a change. The evolution of science and technological change has brought about a number of social responses.

Amidst all of this, there is a need for the policy system to respond to scientific and technological developments – and this response is inextricably linked to related social responses. Even with conventional and predictable developments, it can be difficult to be reflexive and responsive – let alone proactive – in terms of developing policy to support economic change. The challenge is to appropriately consider and mobilize the special places, special processes and special people that underlie the systems.

Policy makers are increasingly pressed to make decisions based on transparent and verifiable evidence in processes that can be validated and audited. The innovation policy file is particularly challenging in that the theory conflicts, the methods are tentative and conclusive evidence is scarce. This chapter critically assesses the body of evidence available to support policy deliberations and offers some signposts for appropriate policy responses.

WHAT DO WE KNOW?

The challenge in all policy debates is to determine what can be accepted as evidence, to assess critically the value of the evidence and to construct what Deborah Stone (1989) calls causal stories that can motivate deliberation, decision and action.

Thomas Kuhn, in *The Structure of Scientific Revolutions* (1970), laid out a remarkable framework that has guided modern thinking about the structure of existing and the staging of new knowledge. This is a good place to start. Essentially, Kuhn's account of the development of scientific knowledge holds that science enjoys periods of stable growth that are periodically interrupted by revisionary surges in methods or revolutionary changes in theory. Kuhn's generalized framework of where new knowledge is created, undergoes change, develops and eventually becomes normalized in society is commonly referenced in discourse around management of knowledge and science. It is particularly relevant in the exploration of cluster activities and the management of science and technology in the agri-food sector, the focus of this book.

According to Kuhn's model, scientific development involves the study of phenomena that lead scientists to develop explanatory theories that, after an appropriate period of challenge and testing, lead to a mature science. If these theories are sustained over time (i.e. through peer review and subsequent trial and error), they become driving paradigms within the scientific system. In terms of trajectory, the evolutionary process involves a period of normal science, where efforts are made to articulate and explore possibilities within the paradigm, where theories are used to predict facts and to solve scientific puzzles, and where new applications of theory are developed. Throughout this process, science is challenged to address both knowns and unknowns. Again, this all takes place within the paradigm of normal science.

A state of normal science is challenged when something exceptional or unknown happens that cannot be accommodated in the paradigm. For the most part, existing theory and methodology usually can adequately and successfully explain new evidence or adapt to more finely articulated causal pathways. At times, however, there may actually be further discoveries of phenomena that violate those expectations that govern normal science. This involves discovery of problems not previously known to have existed and evidence that contradicts the theoretical explanations. At times, normal science cannot fully explain these developments. This can lead to a relatively abrupt shift or transition to a new paradigm. This would require a new conceptual and methodological framework, beyond that of the current normal science.

While Kuhn's examples of the formation and transformation of paradigms are drawn entirely from the history of the physical sciences, the framework can be applied to almost any discipline that seeks to frame itself as science and offer hypotheses and predictions. Both theory and evidence, fundamental features of any science, can be known or unknown. Thus, a classic quadrant box is used to describe the relationships between known and unknown theory and evidence, highlighting four possible situations of science, which have different membership and control mechanisms that can vary in importance depending on the science involved (Table 10.1).

First, normal, conventional or paradigmic science operates where there is a familiar, far-reaching paradigm (with both theory and methodology) that offers predictions that are borne out in the evidence. In these cases, it is possible that new evidence or enhancement of the theories or methodologies could offer finer explanations, but the general conclusions remain constant. Normal science is almost universally accepted by the global academic and scientific community and is stored and shared by those operating in universities and other research centres. This means that normal science is a product of enforced rigour and conformity within the system (including peer review), which reduces the chance for unsubstantiated theory or

Table 10.1. The four realms of science. (Adapted from Phillips, 2007, and Kuhn, 1970.)

	Known impact/evidence	Unknown impact/evidence
Known causes/theory	Normal science (the paradigm)	Hypothetical science
Unknown causes/theory	Experimental science	Speculative science

evidence being perpetuated over long periods. During periods of scientific and technological stability, normal science dominates.

Experimental science is perhaps the most frequent challenge to conventional or normal science. As detection systems become refined (e.g. through new methodologies or data collection methods), it is likely that new evidence may be found that is not explained by the conventional theory. This evidence could challenge theory directly by the discovery of the presence (or absence) of some element that theory says should have not been found (or was not found). It could also challenge the theory indirectly, by illuminating some impact that is not formally incorporated into the theory. This stage of science has historically been very important, as scientists have always sought to discover phenomena that need explaining. The challenge is that, while much experimental work is done in the academic world, an increasing share is done by practitioners or laypersons working in the market and society. These practitioners tend to be isolated from the global scientific community, as they are examining *in situ* events that may only have local, regional or technology-specific impact. The problem with tackling this category of science is that internal governance or control mechanisms are often relatively weak. Measurement and detection methodologies vary widely, depending on what they are designed to address. Evidence may or may not be fully disclosed and often there is no formal peer review structure to assess the significance of new experimental results, either in statistical or scientific terms. This raises the possibility of erroneous measurement of irrelevant evidence that is widely disseminated through the industry or public press or through less formal channels (such as through practitioner newsletters, industry associations and the Internet).

Challenges can also come from the realm of hypothesis-driven science. As new knowledge and techniques accumulate, they raise new questions about the assumptions and structures of normal science. Challenges can range from incremental to fundamental. New hypotheses often emerge when different elements are combined to examine issues on which normal science may be silent. The challenge is that one can never prove any theory or hypothesis – one can only gather evidence that disproves a hypothesis. While most influential hypothetical science is generated within the academic world, which ensures that there is a built-in system for evaluating and testing its relevance, recent trends have complicated this. In the first instance, many fields of science have become a part of Big Science, which has made it more difficult for challenges to surface and be tested solely within the academic scientific community. The price of admission to Big Science is the ability to pair theory with empirical methodologies, which frequently means working with normal science. There are few funders – public or private – that will make substantial investments in highly uncertain science. Meanwhile, there has been a rise in the number of scientists with advanced degrees working outside academia (in industry, government and NGOs), which – paired with increased competition for stories by the media, the emergence of the 'airport' book that exposes the general population to well told causal stories about a specific field (e.g. Richard Florida's *The Rise of the Creative Class*) and some of the more commercially motivated scientific journals – can lead to premature public release of incomplete or often readily refuted hypotheses. Work disseminated through these venues often is not as bound by the rigours of the scientific method as that inside academia. Hence, we often see the promotion of ideas that, upon further consideration and experimentation, can be and are rejected. While new hypotheses are an important part of challenging and advancing knowledge, they must be treated with care as there are probably an infinite number of possibilities to consider.

Finally, there is the realm of speculative science, where there is neither well-articulated theory nor any body of evidence to suggest that a causal relationship is present. This subset of science offers great promise but also poses tremendous challenges. It is probably fair to say that many of the great breakthroughs in science have come from speculation that is not linearly tied to either any specific evidence or to the conventional theories of the day. While this science is important, it most certainly needs to be filtered through further experimentation and theorising before it can provide a base for policy and action. The difficulty is that speculative science is perhaps the most exciting and engaging of all the realms. Hence, it generates significant public scrutiny and debate often before it has been fully tested in the context of experimentation and analysis. The regimes for governing this process are frequently initially undefined; often structures need to be purpose-built to handle review. Academics and practitioners engaging in speculative science generally remain committed to the strictures of peer review but, because speculation is often out of the bounds of normal science, it can be hard to find a suitable forum for full consideration of new ideas. Equally important, many marginally connected academics, non-academic scientists, practitioners and consultants have nothing to lose by speculating – in the parlance of the day, any media coverage is good, as long as they spell your name right.

In the context of innovation, one could make the case that the economic explanation of innovation represents the normal science of the field. The array of economic theories of growth, including both the neoclassical and new growth theories, occupy prime place in the minds of scholars and the key decision makers in treasury departments around the world. But the fact that the theories can only explain a diminishing share of the economic growth observed has laid it open to challenge. The theories, methods, metrics and policy prescriptions that attempt to exploit opportunities involving special places, special processes or special people represent serious challenges to the paradigm. As one might expect with sciences at different stages of evolution and normalization, each has its own range of strengths and weaknesses.

THE PARADIGM – THE ECONOMIC EXPLANATION OF INNOVATION

Arguably, the economic explanation of growth constitutes a particular paradigm that shapes how we understand innovation and its drivers. Economic theories of growth are the oldest and perhaps most well developed of the theories, providing clearly refutable hypotheses that can be tested empirically. Even though it is widely accepted that the theory does not fully explain observed results, it remains at the foundation of most economic development and innovation policy today, if for no other reason than that prices inextricably link inputs and output and are visible signals to analysts.

In the context of the Kuhnian framework, we can see that the economic explanation has a well-developed set of behavioural theories, the almost universal method of economic argumentation (using mathematical logic) and econometric estimation and a large inventory of data that can be used as input and output measures. Alston *et al.* (2000), for example, identified and undertook a meta-analysis of more than 1800 economic estimates of the gains to research in the agri-food space. In short, the economic explanation offers a true paradigmic base for analysis.

The challenge is that its policy prescriptions are somewhat minimalist, there is an accumulation of evidence that does not disprove the theory but rather suggests the theory may not be comprehensive enough, and there is a range of competing or complementary theoretical efforts that explain elements not resolved by the economic theory (some assert these theories are merely solving problems in the paradigm while others assert that they fundamentally challenge the paradigm itself).

Two somewhat contradictory policies are prescribed in the context of the economic explanation. The oldest and most intuitively attractive to non-economists is the infant-industry argument first articulated by Hamilton (1790) and List (1841). Quite simply, it is an economic rationale for trade protectionism. The basic principle behind the argument is that new or budding industries lack the economies of scale to compete with more established competitors and need to be protected until they can attain viable economies of scale. Broadly speaking, protecting infant industries has often involved tariff barriers or discriminatory subsidies. Historically, many now highly industrialized countries successfully industrialized behind tariff barriers (e.g. the USA and Canada). However, these policy measures are now believed not to be the only or even necessarily the most important component in the success in advancing emerging industries. This is especially true today, where there are few examples of innovative firms or sectors that have emerged from protected infant status to be fully competitive.

Since the Second World War, there has been an inexorable expansion of the opinion that open and competitive markets are the best approach to create growing industrial economies. The liberal market consensus now dominates international policy dialogue and is the norm in formal policy debates (Gilpin, 2001). This involves a set of climate-setting measures, including stable and equitable tax and regulatory systems, effective intellectual property rights regimes, and measures and agreements, such as the World Trade Organization, that have leveraged open global markets and are designed to enhance international competition. Generally, then, the burden of proof for further intervention is on the government to demonstrate a market failure and that any compensatory measures will work to address the problem in a manner proportionate to the problem and in a non-discriminatory way. These strictures are actually clearly laid out in the World Trade Organization Agreement of 1995 and now have become judiciable in the context of both trade and domestic policies (especially in the context of the Sanitary and Phytosanitary and Technical Barriers to Trade Sub-Agreements). A second bundle of policy measures related to macro-economic stabilization – e.g. stable and effective financial markets, low inflation, stable exchange rates and sustainable debt-to-GDP ratios – are designed to sustain and support the signalling value of prices. The logic is that policy or market instability leads to uncertainty about prevailing prices, so that savers, borrowers, investors, producers and consumers are unable to detect relative price changes that should translate into shifting efforts that would deliver optimal investment, innovation and economic growth. In short, prices need to reflect market realities and not policy distortions. Finally, while the economic explanation of growth does not explicitly support other measures, most policy agencies advise that open and competitive markets would be enhanced with greater national and international investment in public goods (and activities with significant positive externalities), such as a good primary, secondary and tertiary education system, both public health measures and effective healthcare systems and a range of other justice and welfare policies.

A recent report from the OECD, entitled *The OECD Innovation Strategy: Getting a Head Start on Tomorrow*, sums up the paradigmic policy response. The report asserts that there is no cookie-cutter policy model that governments can implement in support of innovation. Nevertheless, they identify five key priorities for government action that can help to underpin innovation-led development.

First, people should be empowered to innovate. In order to develop and leverage the human and social capital, government policy should concentrate on developing a well-informed and educated citizenry that includes: adapting educational curricula that will prepare prospective students for a rapidly evolving world; improving the quality of teaching through ongoing evaluations and rewards; recognizing that universities play a key role in the innovation enterprise by ensuring the independence, competition, excellence, entrepreneurial spirit, and flexibility in universities; and creating policy regimes to furnish consumers with the capacity to become active contributors in the innovation process and enable them to make informed choices.

Secondly, the innovation capacity in firms must be optimized, as firms are the primary economic conduit for translating ideas and for creating jobs and wealth. The government must: simplify and reduce start-up administrative burdens to reduce the barriers to entry for new firms; structure the labour market policy to recognize that workers need resources in order to be responsive to changes in the economy as firms and industries evolve; facilitate life-long learning; and facilitate finance for business-led innovation. In the event that public funds are used, they should be channelled through existing market-based systems and have a comprehensive market approach.

Thirdly, creating, diffusing and applying new knowledge is vital if firms and countries are to prosper in an increasingly competitive global market. That being said, a number of steps can be taken by government, including: fostering and developing open, free and decentralized communication networks; ensuring that public information remains open so as to allow re-use of innovation; and development of or revisions to patent systems to balance the benefits to innovators with the need to diffuse innovations widely.

Fourthly, innovation can and should be applied to address global and social challenges. Tax policies and other economic instruments can be used as tools to push the market towards innovation. Pricing incentives and disincentives come in explicitly here (e.g. the pricing of environmental externalities such as carbon emissions or the removal of environmentally harmful subsidies). In addition to steering the market towards innovation, the government should provide enough flexibility so that the private sector can pursue the most promising means of addressing global problems.

Finally, the governance and measurement of policies for innovation should continue to be improved. Actions must be coordinated with a clear focus on short-, medium- and long-term goals led by the policy makers at the highest levels. This also includes better communication between different levels of government. Furthermore, improved means of evaluating the innovation process are needed to provide the feedback necessary for policy makers to make informed decisions.

The World Bank's *World Development Report: Reshaping Economic Geography* (2008) (WDR) has added to this list of must-dos with a three-dimensional policy framework that explicitly incorporates factors of density, distance and division. The WDR advocates efforts on three key fronts (what some call the 3-I factors): institutions, infrastructure and incentives. The WDR adds some interesting elements to the OECD package but stops far short of advising governments to engage proactively at the community level or to do much more than set the conditions for innovation processes and creative people to flourish.

The economic paradigm faces significant challenges. At its most fundamental level, the evidence does not conform to the theory or advice derived from the paradigm. Neoclassical growth theory posits that growth should slow absolutely and that economic welfare between advanced and lagging economies should converge, yet neither has happened. While the new endogenous growth theory was an attempt to extend the economic paradigm and solve the problem of purposeful investment in new knowledge to spur growth, the results of both econometric modelling efforts and the practical experience in cities and countries has not (yet) filled the gap. While more refined theory, better methods and more precise measurement may yet sustain the paradigm, in the interim, theorists, analysts and policy advisors are looking for ways to explain the gap or exploit other policy options. A range of new theories has emerged. Inside the paradigm, economic analysts have undertaken major efforts to use longitudinal panel data from more than 180 countries to attempt to determine the importance of the conventional policy variables and to identify potentially undervalued or overlooked causal variables (e.g. Helliwell, 1994). Beyond the paradigm, business theorists, geographers, sociologists and a range of others have constructed theories that consider the specific role of place, processes and people.

EVIDENCE IN SUPPORT OF PLACE-BASED THEORY AND POLICY

Critics argue that the term cluster is vague and has evolved into a form of rhetoric. The critics have a point. The cluster or place-based literature offers a fragmented, partial and contested set of theories, an uneven mix of methodologies and limited comparable evidence to support policy making.

From a scholarly perspective, there is limited consensus even as to how the term cluster is defined, and there appear to be gaps in the literature, particularly related to identifying and testing causal relationships. Most of the work to date simply tells nice stories about places with some evidence of agglomeration. The most glaring weakness is that there has been little or no work on failed clusters. In the absence of counterfactuals, it is risky to draw any definitive conclusions about which features of cluster theory and practice are important and which are trivial. However, it is evident that the notion of clusters has become an important part of the economic vernacular.

There is a strong prima facie case that clusters and firms within clusters benefit from economies of scale by tapping into agglomeration benefits flowing from proximity of competitors and access to support industries and organizations. Broadly speaking, the benefits derived from traded interdependences flow through links between firms, coincidentally enabling the growth of thick labour markets, which offer real value for both people and firms alike. Clusters can and in some circumstances do generate economies of scale, including: access to thick labour markets; physical, transportation or communication infrastructure; sophisticated financing; specialized research institutions and universities; key suppliers or customers; and tailored government policies or programmes.

Economic activity is undoubtedly unevenly distributed – the first precondition for clusters. Industries are almost always more heavily represented in some jurisdictions. Ergo, the cluster. But clusters are more than just simply a geographically proximate collection of firms. Rather, clusters are defined more by the transactions linking them.

Arguably a term of rhetoric, clusters exist in a variety of formats. However, there is no evidence that suggests that a government or community can proactively invent a cluster. The creative spark that incites development of these agglomerations varies from context to context. There are more than 2500 agglomerations of activity in world. Many of these have developed out of sheer luck or by chance and nothing more. While some might appear to be engineered by policy makers, there are enough failed cases to suggest that success is probably neither necessary nor sufficient to conclude that clusters can be designed. As Held (1996, p. 249) points out, 'sadly, in the rush by various governments to employ clusters, some fundamental issues have been slighted, including appropriate research methods and even the definition of the cluster itself'.

That being said, it is possible that clusters can be nurtured and supported. Different policies can have different impacts on capacities of co-located organizations and firms to connect, engage, generate ideas and innovate. Ultimately, firms benefit from cluster-related connections and outcomes. Clusters generate competitive advantage for the member firms in that input costs can be minimized and outputs can be maximized through co-location. In the first instance, clusters may work to lower both transactional costs and risk, first by offering more choice of suppliers and buyers, and secondly by offering secondary markets to recycle and repurpose inefficiently allocated labour and capital. Moreover, clusters appear to reduce the uncertainty of long-term economic growth by enabling firms to strategically identify, generate and adapt to local and global technological progress.

It would be fair to assume that knowledge-based innovation clusters may have compressed generations in their life cycle compared with old-style industrial clusters. Given the research focus in Saskatoon (and in many of the other biotechnology clusters around the world), many are asking whether such a cluster can ever be self-sustaining. Clusters that depend on research and development are vulnerable to volatile research priorities. The challenge, then, is to determine whether the current base is sustainable, whether a science and technology-based cluster can extend beyond its research base and whether a cluster offers any potential to add value either up- or downstream to input and output sectors (via capture of logistics, value capture and brand management).

The Saskatoon agricultural biotechnology cluster examined in detail in Chapter 4 offers some insights into the workings of clusters, especially in the global agri-food space. In the first instance, it appears to follow some form of life cycle. In a Kuhnian fashion, the cluster has evolved over time, enjoying periods of growth punctuated by surges of new knowledge development. When technology is emerging or transformative, firms tend to gather together and collaborate in order to share tacit and non-codified knowledge. As the industry gains momentum, knowledge becomes more institutionalized (i.e. patented) and systems are developed to maximize profits and minimize costs, there seems to be less reason to collaborate or to use open business models. This reduces the incentives for agglomeration. In fact, as technology in Saskatoon evolved and products came to market, some of the large MNEs (such as Bayer, BASF and Monsanto) either moved away or downsized their local operations. The need to be proximate (or at least that proximate) diminished as time went on.

This approach suggests that knowledge-based innovation systems are inherently unstable and vulnerable to the vagaries of research. In this sense, they are unlikely ever to be truly self-sustaining. Innovation clusters probably go through cycles, with periods where they require high reinvestment (public, private or both) and periods of high pay-out. Davis (2003) offered a five-stage life-cycle model for knowledge clusters, based on his observations of the ICT cluster in New Brunswick. The model appears to operate in Saskatoon. The first stage includes the accumulation of assets, which may take place over a number of years. In many instances, these investments come from the public sector (e.g. the recent infusion to genomics research). Secondly, assets are converted into business resources through firm entrepreneurial effort. Thirdly, firms grow by exploiting new resources and capabilities in external markets. Fourthly, as production and markets mature, collective efficiencies are realized. Finally, various market and non-market selection processes lead to local specialization. This life-cycle model suggests that, if innovation is going to continue, it will require regular reinvestment in the accumulation of assets. This can come from recycling returns on investments or from outside infusions. If the innovations resulting from the cluster are commercialized by locally based firms, there is a reasonable prospect that this reinvestment may come about naturally (e.g. in Montreal, St Louis or in Research Triangle Park, North Carolina). In contrast, clusters dependent on foreign multinationals to commercialize innovative products or services are likely to see outflows of the returns to the innovation, such that there may not be enough capital within the community to fully reinvest to restart the process.

Although theoretically possible, there is no clear evidence from Saskatoon that clusters can facilitate diversification into other related areas such as bioenergy, nutraceuticals or biomedicine. While intuitively it is conceivable that there are synergies between similar sectors (such as the bio-based fields), the corporate effort to realize these synergies in the 1990s through life-science mergers among the MNEs was largely unsuccessful. Nevertheless, what might not be possible within companies may still be possible within regions. Procyshyn (2004) examined the potential for the Saskatoon cluster to diversify into new applications, in response to recent investments in the CLS synchrotron. This work looked at the functional requirements of the innovative agro-biotechnology sector and assessed the potential for the central actors providing those functions to extend and expand their offerings to the larger life-science community. Theoretically, there is an argument that this may be possible, but no experimental or empirical evidence is available. It is likely that Saskatoon will be a test case, as the recent opening of the CLS offers the local innovation system an opportunity to adapt or for a new system to evolve. While Ryan (2011) and Ryan *et al.* (2012) have detected the emergence of science and research networks around the facility, there is no evidence yet that this is translating or will translate into more commercialization.

While much of the work in this book has looked at the regional level, it is fair to inquire what it says about innovation beyond the lead sector or the region. There is no compelling evidence from the examination of the mining and ag-biotechnology examples in Saskatoon that new unrelated clusters can be spawned in the same region. Moreover, there is little or no theoretical or empirical support for the assertion that a successful cluster in one region will have any causal influence on the emergence or evolution of new clusters in other regions. As a start, Phillips *et al.* (2004) assessed seven biotechnology-based innovation systems across Canada but found no evidence of any significant links. Perhaps the most that can be expected is that a successful cluster will provide demonstration value – or inspiration – to actors in other sectors and communities.

Knowledge-based development is inherently different from traditional industrial development. While infant industry protection (discussed above) made some sense in the industrial context, it is not clear whether it has any value in a knowledge-based world. The *entrepôt* model of innovation poses some serious challenges for development policy. Much of current development effort has a strong mercantilist orientation, with a focus on self-sufficiency. Governments at all levels in many countries are keen to use their tax and fiscal policy to encourage greater local R&D or to attract global firms to relocate their R&D programmes into their jurisdiction in an effort to generate higher value exports or to replace imports. This often involves preferential support for national champions or exclusive deals for relocating MNEs. Usually governments do this without any consideration of the corresponding relationships and interactions that knowledge-based firms require to succeed. If innovation can be thought of as limited to within a firm or within a regional or national community, then such a narrow approach might have some chance of succeeding. But if innovation is global, as appears to be the case in the life sciences, then narrow, mechanistic self-sufficiency strategies may either fail or at times prove to be counterproductive. The evidence from the biotechnology sector suggests that innovation is truly global, which goes a long way towards explaining the evidence in Chapters 6 and 8 that both firms and skilled employees are more interested in the innovation community than in fiscal incentives, public infrastructure or other government support. By extension, a mercantilist policy that discourages global links could not only fail to attract but could ultimately drive out local firms or researchers as they seek access to the global community.

EVIDENCE IN SUPPORT OF INNOVATION SYSTEMS

Innovation systems hold significant fascination. Lundvall (1992) posited that national innovations are the proper unit of analysis (albeit based on work in the small, unitary and relatively homogenous Scandinavian states), while Gertler *et al.* (2000) argued for regional systems of innovation (based on their work in Canada, a large, continental economy) and Etzkowitz and Leydesdorff (2000) focused on localized triple-helix relationships between government, industry and universities. While their reference points vary, each approach is characterized by their focus on economies of scope and untraded interdependencies. Social capital is a key part of the *lingua franca* in this literature. However, when drilling down into knowledge-based theory and practice, Marshall's (1890) characterization of 'mysteries in the air' appears to describe most aptly the phenomenon scholars are seeking to define and study.

The theory and evidence focused on untraded interdependencies is problematic. While there has been substantial work on framing social systems and processes related to innovation, there is a range of competing and sometimes conflicting theoretical approaches. If looked at from an appropriate distance, one might conclude that the different scales of analysis chosen – such as geographical and institutional – are simply applications of a broader and more universal theory. One reading of the theory would support this view, as most of the literature is focused on defining and examining some form of social capital and how it works to mobilize people and institutions in risky innovative ventures. The challenge is that the underlying behavioural assumptions are not always consistent. Some of the literature implies that people engaged in creating and exploiting social capital are probably acting as rational actors, using either implicit or explicit shadow prices to mediate their efforts. Other parts of the theory see this in a much more social context, with any economic impact an unintended by-product of the processes. Perhaps more importantly for policy makers, the theory has not convincingly been validated through any standardized methods or generated any substantial body of empirical evidence. Much of this work is either conceptual or case based. The Oslo and Frascati manuals are a start in developing a common base of evidence, but to date the surveys offer only glimpses of what must undoubtedly be happening in firms and communities if the theory is correct.

The constant ebbs and flows and evolving knowledge-based economic environment make it difficult to pin down processes or to offer definitive policy prescriptions in the innovation systems space. In the absence of clear signals, governments are advised and usually undertake to support universities and other post-secondary institutions as foundational parts of the knowledge system. Teaching and research carried out in these institutions (and some specialized research laboratories) is assumed to generate local capacity to absorb knowledge, technologies or commercial ventures from other competing centres. The institutions also tend to trigger significant investments in research infrastructure and in programmes to sustain research. Grants and monetary incentives are also needed to support targeted research projects, to encourage collaborations or joint ventures and to facilitate technology transfer. Intellectual property regimes – to a greater or lesser degree – are also useful in ensuring the involvement of economically motivated individuals and private firms.

Many different metrics have been developed to measure innovation, but more work is needed. Many of the measures used to examine the innovation process are input measures which fail to correlate well with the output metrics used to measure innovation. The continued disconnect within the innovation systems literature between input and output measures makes it difficult to measure the outcomes and impacts of government policy on innovation and, as a result, to provide policy recommendations.

Over the last decade, governments around the world have financially sponsored programmes and projects that foster partnerships and encourage collaborations across disciplines and national boundaries. At root is a relatively untested assumption that larger research ventures with more diverse knowledge bases improve the potential for re-combinations that will accelerate invention and innovation. Public financial support is rendered to projects or ventures that contribute to the nation's vision and priorities. A large number of developed and transitional economies are increasingly implementing projects with a wider scale and scope in order to extend their reach and realize any scope benefits. The state in its public venture capital role facilitates Big Science initiatives and is looking at ways to overcome the shortage of private investments for start-ups and early stage projects due to their uncertain profit margins. The financial support for new initiatives is usually accompanied and directed by customized science and technology policy frameworks and agendas (Table 10.2).

One key to succeeding in this type of a world would be to invest in those institutions and mechanisms that generate the capacity for the economy to absorb science, technology and innovation from the global pipelines. The literature and studies in Saskatoon and elsewhere offer insights into a number of vital elements to creating capacity (Table 10.3). First, there must be effective mechanism to both practically and legally transfer knowledge, which at a minimum involves a domestic research community with international collaborations, some MNEs with proprietary technologies and appropriate legal protection for intellectual property. Secondly, there needs to be open

Table 10.2. OECD member state STI policy frameworks. (From http://www.oecd.org/infobycountry/0,3380,en_2649_33703_1_1_1_1_1,00.html.)

Country	S&T policy frameworks	Common network encouraging strategic elements across national science policy frameworks
Australia	Powering Ideas, 2009	Strengthen integrated approach to innovation and improve Australia's linkages with global innovation systems
Canada	Mobilizing Science and Technology to Canada's Advantage, 2007	One of the core strategic principles is 'fostering partnership'
Denmark	Globalisation Strategy, 2012	Focus on efforts that contribute to networking and collaboration with worldwide research initiatives
Finland	Innovation Strategy, 2008	Encourage key stakeholder involvement in the innovation process and in the development of collaborative alliances amongst domestic firms involved in innovation activities
France	National Research and Innovation Strategy, 2008	Prioritize synergized innovation efforts among stakeholders present in competing innovation clusters
Germany	High-tech Strategy, 2020	Encourage innovation-based linkages
Netherlands	R&D Promotion Act (WBSO), 1994, 2005	Add funds to strengthen domestic and foreign innovation linkages
Sweden	Research and Innovation Bill, 2008	Renew funding to promote sustained research relationships
UK	Science and Innovation Investment Framework (SIIF), 2004–2014	Focus research and innovation activities on large innovative firms and strong internal or foreign linkages
USA	American Recovery and Reinvestment Act, 2009	Allocate financial backing to large-scale partnership oriented innovation models

and accessible labour markets for skilled workers. Third, there needs to be a platform – e.g. the public laboratories in the NRC-PBI and AAFC in Saskatoon or industry-led networks – for community-based synergies to develop. These elements provide the foundation for absorbing global knowledge. Others,

Table 10.3. Public policy options for nurturing knowledge-based innovation clusters.

Knowledge factor	Policy prescription
Know-why	Develop absorptive capacity through basic research capacity in universities and public laboratories; nurture two-way international flows of information through programmes that support and encourage international collaborations
Know-what	Create effective intellectual property protection systems that facilitate two-way international flow of innovations; encourage location of competing and collaborating multinational enterprises to encourage transfer of proprietary technologies
Know-how	Develop and maintain a critical mass of researchers and technologists focused on advancing and using the technologies either in public laboratories (e.g. AAFC) or in private firms; nurture thick labour markets through post-doctoral research support and liberal labour policies
Know-who	Develop and maintain open-platform institutions to facilitate research collaborations (e.g. NRC); nurture collaborative, industry-led networks to facilitate communications and to assist with developing the forward and backward institutions necessary for efficient commercialization

such as preferential financing and speciality commercial services may be important, but would appear to be second-order requirements.

In short, innovation clusters are attractive economic development tools, but they must be nurtured with an appreciation for their partial and incomplete nature. Fundamentally, they are part of a global innovation system, and cannot thrive if cut off from the lifeblood of the system – ideas, skilled labour and collaborative knowledge platforms.

EVIDENCE IN SUPPORT OF CREATIVE CITIES

The most attractive and exciting option facing policy makers is to nurture creative talent in creative cities. The allure is undeniable – not only do policy makers and civic leaders get the promise of spurring cutting-edge innovation, they also have the prospect of engaging with and personally benefiting from working with and living beside exciting and attractive individuals. Alas, there is no magic elixir. While some communities may indeed gain from such policies and initiatives, most communities in the world appear to have limited potential.

People are undoubtedly at the root of any innovation. Firms, universities and social processes are fundamentally about groups of people. The question is what kind of people generate the greatest impact. The literature is mixed. Some of the literature focuses on the characteristics of specific individuals – be they entrepreneurs, creatives, leaders, talent or stars – and seeks to understand how these exciting people translate their energies into innovation outcomes. The human and social capital that these people accumulate and use as labourers, knowledge generators and members of the community is vital – but hard to measure and even harder to exploit.

The challenge is that both theory and evidence of the creative cities hypothesis has lagged the enthusiasm of practitioners and pundits. The theory for why creative, talented people might move and then how they would specifically contribute to economic development and innovation is intuitive but not overly well developed. Moreover, there is empirical work showing that relationships may exist, but the results are at best tentative and in some ways are disputed by counterfactuals.

There is some evidence that research stars, in particular, play a key role in knowledge generation in research-based economies in that they are responsible for higher than average levels of output in terms of patents or publications and in many ways are key actors in translating the tacit and codified knowledge generated in universities and research laboratories into commercial use (Zucker *et al.*, 1998). The evidence is less clear about whether one can proactively manage stars. Evidence from the USA and even from case work in Saskatoon shows that, while sustained public investment in specific research areas may be necessary for stars to emerge, it is less clear that one can proactively recruit stars to kick start this process. The work in US biotechnology and the studies in Canada can identify stars *ex post*, but there are no easy ways to identify and recruit potential stars early in their career. Furthermore, even if one could recruit full-blown stars, there is no convincing evidence that if they are moved to a new venue they would be able to fully effect knowledge translation. In many ways, the exchange of knowledge through entrepreneurial ventures is probably as much a result of deep roots and wide networks as the unique skills and capacities that individual stars might have.

Florida's creative cities hypothesis is more problematic. Creative talent – in his work characterized more as Bohemian, cultured artisans rather than innovative boffins and entrepreneurs – undoubtedly agglomerate and cluster, but one cannot properly determine whether their agglomeration is cause or effect. So far, most of the work has involved correlation analysis, which cannot discern causality. Some assert that creative, mobile people are attracted by tolerant communities endowed with appropriate technology and cultural infrastructure, and that their agglomerations then generate innovation and economic development. An alternative hypothesis is that economic development generates the necessary scale to attract a critical mass of creative types, who then go on to generate innovative outcomes. If the former, then policy makers may be able to invest in social engineering and community infrastructure – on the assumption that once an appropriate venue is built, creatives will come. If, on the other hand, creative communities are simply a by-product of other development processes, then proactively investing in cultural amenities may be premature or may actually depress other innovative processes, as it draws economic and social capital away from more valuable ventures.

Our work in Saskatoon looking primarily at the agri-food system offers some support for the important role of research stars but no real support for the creative cities hypothesis. Interestingly, recent work in Montreal and Calgary, both undeniably larger and more cosmopolitan cities with aspirations of being cultural havens, suggests that even there talented people are primarily attracted by economic opportunity and not by the cultural amenities of the centre (Darchen and Tremblay, 2011; Ryan *et al.*, 2012).

It is undoubtedly true that people are mobile and that when they move they bring important new ideas and connections. Thus, whenever we attract and integrate economic and cultural migrants we can, in theory, enhance the potential for innovation. All of this is intuitive. The concern with the creative city hypothesis is that it assumes that one can socially engineer this to happen. We doubt that this will work for many communities, especially centres that are primarily engaged in agricultural or extractive industries. The OECD offers a fall-back strategy. They suggest that governments everywhere should focus on human capital policies rather than proactive investment in cultural amenities. They suggest that governments pursue and implement policies that help to nurture existing high-quality personnel (such as life-long learning and exchange programmes) and encourage mobility and labour market flexibility. Even if these policies do not generate the creative city phenomena, they conform to the advice that flows from the economic paradigm. In that sense, they cannot hurt and might help.

CONCLUSION

Despite our best efforts in this book to offer unambiguous and compelling policy advice for those concerned about innovation in the bio-based economy, we are forced to conclude that creative tensions exist. Achieving economic success, from a regional perspective, requires a balance between good management practice (policies and firm-based strategies) and letting the cards fall where they may. It involves continued attentiveness to the tenuous relationship between the state and the market, the broader implications of open versus proprietary research, and the balance between horizontal and vertical strategies. Rullani (2002, p. 51) asserted that innovation communities are living systems, 'organism[s] made of past experiences, collective identities and relationships'. This characterization goes beyond the narrow economic ideal. It neither readily fits the economic theories nor with the more contextual theories that focus on place, processes and people. People, knowledge, institutions and a variety of other elements come together in serendipitous ways to create this complex phenomenon. Thinking about innovation requires awareness of not only the incomplete and partial character of communities and systems but also of the dynamics of process and change. This fundamentally forces us to think about the role of distributed control mechanisms, emergence, re-emergence and self-organization, the ebb and flow of actors and agents, and the continual adjustment of behaviours and attitudes. In this context, history matters – a lot.

In the end, innovation undoubtedly requires a tailored use of differentiated policy instruments to address integration challenges that range from the relatively straightforward to the most complicated.

References

Acemoglu, D. and Robinson, J. (2008) *The Role of Institutions in Growth and Development.* Commission on Growth and Development, Working Paper No. 10. The World Bank, Washington, DC <http://www.growthcommission.org/storage/cgdev/documents/gc-wp-010_web.pdf>.

Alston, J., Mara, M., Pardey, P. and Wyatt, T. (2000) Research returns redux: a meta-analysis of the returns to agricultural R&D. *Australian Journal of Agricultural and Resource Economics* 44(2), 185–215.

Archibugi, D., Howells, J. and Michie, J. (eds) (1999) *Innovation Policy in a Global Economy.* Cambridge University Press, Cambridge.

Arrow, K. (1962) *Economic Welfare and the Allocation of Resources for Invention in Rate and Direction of Inventive Activity.* NBER and Princeton University Press, Princeton, New Jersey, pp. 609–625.

Arthur, B., Durlauf, S. and Lane, D. (eds) (1997) *The Economy as an Evolving Complex System II.* Addison Wesley, Reading, Massachusetts.

Asheim, B. and Coenen, L. (2005) Knowledge bases and regional innovation systems: comparing nordic clusters. *Research Policy* 34, 1173–1190.

Asheim, B. and Gertler, M. (2005) The geography of innovation: regional innovation systems. In: Fagerberg, J., Mowery, D. and Nelson, R. (eds) *The Oxford Handbook of Innovation.* Oxford University Press, Oxford, pp. 291–317.

Asheim, B., Isaksen, A., Nauwelaers, C. and Tödling, F. (eds) (2003) *Regional Innovation Policy for Small-Medium Enterprises.* Edward Elgar Publishing, Cheltenham, UK/Northampton, Massachusetts.

Atlantic Canada Opportunities Agency (ACOA) (1997) *Prospects for Growing Knowledge-based Industrial Clusters in Atlantic Canada.* Government of Canada: Atlantic Canada Opportunities Agency <http://www.acoa.ca/e/library/reports/knowledge_industrial.shtml>.

Audretsch, D. (1998) Agglomeration and the location of innovative activity. *Oxford Review of Economic Policy* 14, 18–29.

Audretsch, D. (2002) The innovative advantage of US cities. *European Planning Studies* 10, 165–176.

Audretsch, D. and Feldman, M. (1996) R&D spillovers and the geography of innovation and production. *American Economic Review* 86, 630–640.

Bachrach, P. and Baratz, M. (1962) Two faces of power. *American Political Science Review* 56, 947–952.

Baldwin, R. and Krugman, P. (1988) Industrial policy and international competition in wide-bodied jet aircraft. In: Baldwin, R. (ed.) *Trade Policy Issues and Empirical Analysis*. National Bureau of Economic Research Inc., Cambridge, Massachusetts, pp. 45–78.

Bannock, G., Baxter, R. and Rees, R. (1972) *The Penguin Dictionary of Economics*. Penguin Books, London.

Barney, J. (1995) Looking inside for competitive advantage. *Academy of Management Executive* 9, 49–61.

Barro, R. (1991) Economic growth in a cross section of countries. *Quarterly Journal of Economics* 106, 407–443.

Bathelt, H., Malmberg, A. and Maskell, P. (2004) Clusters and knowledge: local buzz, global pipelines and the process of knowledge creation. *Progress in Human Geography* 28, 31–56.

Beinhocker, E. (2006) *Origin of Wealth: Evolution, Complexity and the Radical Remaking of Economics*. Harvard Business School Press, Boston, Massachusetts.

Bekele, G. and Jackson, R. (2006) *Theoretical Perspectives on Industry Clusters*. Regional Research Institute, West Virginia University, Research Paper 2006–5 <http://www.rri.wvu.edu/pdffiles/bekelewp2006-5.pdf>.

Belleflamme, P., Picard, P. and Thisse, J.-F. (2000) An economic theory of regional clusters. *Journal of Urban Economics* 48, 158–184.

Benkler, Y. (2006) *The Wealth of Networks: How Social Production Transforms Markets and Freedom*. Yale University Press, New Haven, Connecticut.

Bergman, E. and Feser, E. (1999) Industry clusters: a methodology and framework for regional development policy in the United States. In: OECD Proceedings, *Boosting Innovation: The Cluster Approach*. OECD, Paris.

Bergman, E. and Feser, E. (2004) *Industrial and Regional Clusters: Concepts and Comparative Applications*. The Web Book of Regional Science, West Virginia University <http://www.rri.wvu.edu/WebBook/Bergman-Feser/contents.htm>.

Bolan, M. (1997) The mobility experience and neighborhood attachment. *Demography* 34, 225–237.

Boland, W., Phillips, P. and Ryan, C. (2010) Centerless governance and the management of global R&D: public–private partnerships and plant-genetic resource management. In: *Proceedings of the VIII Triple Helix Conference*, Madrid, 20–22 October.

Bolton, B. and Thompson, J. (2004) *Entrepreneurs: Talent, Temperament, Technique*, 2nd edn. Elsevier Butterworth-Heinemann, Oxford, UK.

Bonacich, P. (1972) Factoring and weighing approaches to status scores and clique identification. *Journal of Mathematical Sociology* 2, 113–120.

Boorstein, D. (1983) *The Discoverers*. Random House, New York.

Borgatti, S., Everett, M.G. and Freeman, L.C. (2002) UCINet for Windows: Software for Social Network Analysis. Analytic Technologies, Cambridge, Massachusetts.

Boschma, R. and Frenken, K. (2006) Why is economic geography not an evolutionary science? Towards an evolutionary economic geography. *Journal of Economic Geography* 6, 273–302.

Boschma, R. and Martin, R. (2007) Editorial: Constructing an evolutionary economic geography. *Journal of Economic Geography* 7, 537–548.

Boschma, R. and Martin, R. (2010) The aims and scope of evolutionary economic geography. *Papers in Evolutionary Economic Geography (PEEG)* 10.01, 1–46.

Boulding, K. (1973) *The Economy of Love and Fear.* Wadsworth, Belmont, California.

Bourdieu, P. (1985) The social space and the genesis of groups. *Theory and Society* 14, 723–744.

Bourdieu, P. (2006) The forms of capital. In: Lauder, H., Brown, P., Dillabough, J.-A. and Halsey, A.H. (eds) *Education, Globalization, and Social Change*, Oxford University Press, New York.

Boxman, E., de Graaf, P. and Flap, H. (1991) The impact of social and human capital on the income attainment of Dutch managers. *Social Networks* 13, 51–73.

Brander, J. and Spencer, B. (1985) Export subsidies and international market share rivalry. *Journal of International Economics* 18, 83–100.

Braudel, F. (1973) *Capitalism and Material Life, 1400–1800.* Translated from French by Miriam Kochan. Harper and Row, New York.

Brink, J., McKelvey, M. and Smith, K. (2004) Conceptualizing and measuring modern biotechnology. In: McKelvey, M., Rickne, A. and Laage-Hellman, J. (eds) *The Economic Dynamics of Modern Biotechnology*, Edward Elgar, Cheltenham, UK, pp. 20–42.

Brown, J. and Duguid, P. (2000) *The Social Life of Information.* Harvard Business School Press, Boston, Massachusetts.

Buenstorf, G. and Fornahl, D. (2006) *B2C – Bubble to Cluster: The Dot.com Boom, Spin-off Entrepreneurship, and Regional Industry Evolution.* Papers on Economics and Evolution 2006–20. Max Planck Institute of Economics, Evolutionary Economics Group, Jena, Germany.

Buenstorf, G. and Klepper, S. (2005) *Heritage and Agglomeration: The Akron Tire Cluster Revisited.* Papers on Economics and Evolution 2005–08. Max Planck Institute of Economics, Evolutionary Economics Group, Jena, Germany.

Burns, J. (1978) *Leadership.* Harper & Row, New York.

Burt, R. (1992) *Structural Holes: the Social Structure of Competition.* Harvard University Press, Cambridge, Massachusetts.

Burt, R. (2000) The network structure of social capital. *Research in Organisational Behaviour* 22, 345–423.

Burt, R. (2005) *Brokerage and Closure: an Introduction to Social Capital.* Oxford University Press, New York <http://gsbwww.uchicago.edu/fac/ronald.burt/research/>.

Carley, K. (1986) Knowledge acquisition as a social phenomenon. *Instructional Science* 14, 381–438.

Carley, K. and Reminga, J. (2004) *ORA: Organization Risk Analyzer.* Center for Computational Analysis of Social and Organizational Systems <http://reports-archive.adm.cs.cmu.edu/anon/isri2004/CMU-ISRI-04-106.pdf>.

Castells, M. (2000) *The Rise of the Network Society*, 2nd edn. Blackwell Publishers, Oxford, UK.

Christopherson, S. and Rightor, N. (2010) The creative economy as 'Big Business'. *Journal of Planning Education and Research* (special issue on *The Creative Economy and Economic Development*) 29, 336–352.

Clayman, B. and Holbrook, J. (2004) *Surviving Spin-offs as a Measure of Research Funding Effectiveness.* Centre for Policy Research on Science and Technology (CPROST), paper no. 04-03. <http://www.sfu.ca/cprost/reports.html>.

Clifton, D. and Harding, R. (1986) *A Statistical Analysis of the Psychometric Properties of the SRI Entrepreneur Interview.* Gallup Report, September.

Coase, R. (1937) The nature of the firm. *Economica* 4, 386–405.

Coase, R. (1960) The problem of social cost. *Journal of Law and Economics* 3, 1–44.

Coe, D. and Helpman, E. (1995) *International R&D Spillovers.* NBER Working Paper no. 4444. National Bureau of Economic Research, Cambridge, Massachusetts.

Coe, D., Helpman, E. and Hoffmaister, A. (1997) North–south R&D spillovers. *Economic Journal* 107, 134–149.

Coenen, L. and Asheim, B. (2004) The role of regional innovation systems in a globalising economy: comparing knowledge bases and institutional frameworks of Nordic clusters. In: *DRUID Summer Conference*, Elsinore, Denmark <http://www.druid.dk/conferences/summer2004/papers/ds2004-166.pdf>.

Coenen, L., Moodysson, J. and Asheim, B. (2004) Nodes, networks and proximities: on the knowledge dynamics of the Medicon Valley biotech cluster. *European Planning Studies* 12, 1003–1018.

Coenen, L., Moodysson, J., Ryan, C., Asheim, B. and Phillips, P. (2006) Comparing a pharmaceutical and an agro-food bioregion: on the importance of knowledge bases and spatial patterns of collaboration. *Industry and Innovation* 13, 393–414.

Coleman, J. (1990) *Foundations of Social Theory.* Belknap Press of Harvard University Press, Cambridge, Massachusetts.

Committee for Study of Invention (2004) *Invention: Enhancing Inventiveness for Quality of Life, Competitiveness, and Sustainability.* Report of the Committee for Study of Invention, sponsored by the Lemelson-MIT Program and the National Science Foundation <http://web.mit.edu/invent/n-pressreleases/downloads/report.pdf>.

Cooke, P. (1992) Regional innovation systems: competitive regulation in the new Europe. *Geoforum* 23, 365–382.

Cooke, P. (1998) Introduction: origins of the concept. In: Braczyk, H., Cooke, P. and Heidenreich, M. (eds) *Regional Innovation Systems*, 1st edn. UCL Press, London.

Cooke, P. (2001) Regional innovation systems, clusters, and the knowledge economy. *Industrial and Corporate Change* 10, 945–974.

Cooke, P. (2002) *Knowledge Economies: Clusters, Learning and Co-operative Advantage.* Routledge, New York.

Cooke, P., Roper, S. and Wylie, P. (2003) 'The golden thread of innovation' and Northern Ireland's evolving regional innovation system. *Regional Studies* 37, 365–379.

Dahl, R. (1961/1989) *Who Governs? Democracy and Power in an American City.* Yale University Press, New Haven, Connecticut.

Darchen, S. and Tremblay, D. (2011) *The Attraction/Retention of Knowledge Workers and the Creative City Paradigm: Can We Plan for the Talents and at What Cost? The Case of Montreal.* Research Note of the Canada Research Chair on the Socio-organizational Challenges of the Knowledge Economy. Téluq-UQAM.

Davis, C. (2003) *Growth Drivers of a Regional ICT Cluster: the Case of New Brunswick.* Canadian–Nordic Workshop on Clusters and Regional Innovation Systems, Gotenberg, Sweden.

Dobni, B. and Phillips, P. (2001) *The Saskatoon Science Map™.* University of Saskatchewan, Saskatoon, Canada.

Doeringer, P. and Terkla, D. (1995) Business strategy and cross-industry clusters. *Economic Development Quarterly* 9, 225–237.

Doloreux, D. (2004) Regional innovation systems in Canada: a comparative study. *Regional Studies* 38, 479–492.

Doloreux, D. and Parto, S. (2005) Regional innovations systems: current discourse and unresolved issues. *Technology in Society* 27, 133–153.

Dosi, G. (1984) *Technical Change and Industrial Transformation: the Theory and an Application to the Semiconductor Industry.* Macmillan Press, London.

Drucker, P. (1993) *Post-Capitalist Society.* Harper Business, New York.

Drucker, P. (1999) *Management Challenges for the 21st Century.* Harper Business, New York.

Dunning, J. (2001) *Global Capitalism at Bay?* Routledge, London.

Edquist, C. (ed.) (1997) *Systems of Innovation: Technologies, Institutions and Organisations.* Pinter, London.

Edquist, C. and Johnson, B. (1997) Institutions and organizations in systems of innovation. In: Edquist, C. (ed.) *Systems of Innovation: Technologies, Institutions and Organisations.* Pinter, London, UK, pp. 41–63.

Endacott, G. (1964) *An Eastern Entrepot: a Collection of Documents Illustrating the History of Hong Kong Overseas Research Publication 4.* HM Stationery Office, London.

Enright, M. (1993) The geographic scope of competitive advantage. In: Dirven, E., Groenewegen, J. and van Hoof, S. (eds) *Stuck in the Region? Changing Scales of Regional Identity.* Netherlands Geographical Studies, Utrecht, The Netherlands, pp. 87–102.

Enright, M. (1998) Regional clusters and firm strategy. In: Chandler, A., Hagstrom, P. and Solvell, O. (eds) *The Dynamic Firm: the Role of Technology, Strategy, Organization and Regions.* Oxford University Press, Oxford.

Enright, M. (2000) Regional clusters and multinational enterprises: independence, dependence, or interdependence? *International Studies of Management & Organization* 30, 114–138.

Etzkowitz, H. and Leydesdorff, L. (1995) The triple helix: university–industry–government relations: a laboratory for knowledge based economic development. *European Society for the Study of Science and Technology Review* 14, 14–19.

Etzkowitz, H. and Leydesdorff, L. (2000) The dynamics of innovation: from national systems and mode 2 to a triple helix of university–industry–government relations. *Research Policy* 29, 109–123.

Etzkowitz, H. and Leydesdorff, L. (2011) *The Triple Helix of Innovation* <http://www.leydesdorff.net/spp98/>.

Etzkowitz, H. and Ranga, M. (2010) A triple helix system for knowledge-based regional development: from 'spheres' to 'spaces'. In: *Proceedings of the VIII Triple Helix Conference*, Madrid, October.

European Association of Development Agencies (2004) *Can Regions Make the Difference?* <http://www.eurada.org/files/SME%20support/Can%20regions%20make%20the%20difference_E.pdf>.

Florida, R. (2002a) Bohemia and economic geography. *Journal of Economic Geography* 2, 55–71.

Florida, R. (2002b) *The Rise of the Creative Class*. Basic Books, New York.

Florida, R. (2002c) The economic geography of talent. *Annals of the Association of American Geographers* 92, 743–755.

Florida, R. (2005) The world is spiky: globalization has changed the economic playing field, but hasn't levelled it. *Atlantic Monthly*, October.

Freeman, C. (1987) *Technology and Economic Performance: Lessons from Japan*. Pinter Publishers, London.

Freeman, C. (1995) The national system of innovation in historical perspective. *Cambridge Journal of Economics* 19, 5–24.

Fukuyama, F. (1996) *Trust*. Penguin Books, New York.

Gassmann, O. and Gaso, B. (2004) Insourcing creativity with listening posts in decentralized firms. *Creativity and Innovation Management* 13, 3–14.

Gault, F. (2010) *Innovation Strategies for a Global Economy: Development, Implementation, Measurement and Management*. Edward Elgar, Cheltenham, UK.

Gertler, M. and Levitte, Y. (2005) Local nodes in global networks: the geography of knowledge flows in biotechnology innovation. *Industry and Innovation* 13, 487–507.

Gertler, M., Wolfe, D. and Garkut, D. (2000) No place like home? The embeddedness of innovation in a regional economy. *Review of International Political Economy* 7, 688–718.

Gertler, M., Florida, R., Gates, G. and Vinodrai, T. (2002) *Competing on Creativity: Placing Ontario's Cities in Continental Context*. Institute for Competitiveness and Prosperity and the Ontario Ministry of Enterprise, Opportunity and Innovation, Toronto, Canada.

Gibbons, M., Limoges, C., Nowotny, H., Schwartzman, S., Scott, P. and Trow, M. (1994) *The New Production of Knowledge: the Dynamics of Science and Research in Contemporary Societies*. Sage Publications, London.

Gilpin, R. (2001) *Global Political Economy: Understanding the International Economic Order*. Princeton University Press, Princeton, New Jersey.

Gladwell, M. (2000) *The Tipping Point*. Little, Brown and Co., New York.

Gladwell, M. (2008) *Outliers*. Little, Brown and Co., New York.

Glaeser, E. (2004) *Review of Richard Florida's* The Rise of the Creative Class. <http://creativeclass.com/rfcgdb/articles/GlaeserReview.pdf>.

Godelier, M. (1999) *The Enigma of the Gift*. Polity Press and the University of Chicago, Chicago, Illinois.

Gollub, J.O. (1997) *Cluster-based Economic Development: a Key to Regional Competitiveness*. Economic Development Administration, US Department of Commerce, Washington, DC.

Gordon, I. and McCann, P. (2000) Industrial clusters: complexes, agglomeration and/or social networks? *Urban Studies* 37, 513–532.

Granovetter, M. (1983) The strength of weak ties: a network theory revisited. *Sociological Theory* 1, 201–233.

Granovetter, M. (1985) Economic action and social structures: the problem of embeddedness. *American Journal of Sociology* 91, 481–510.

Grant, R. (2002) The knowledge-based view of the firm. In: Choo, C. and Bontis, N. (eds) *The Strategic Management of Intellectual Capital and Organizational Knowledge*. Oxford University Press, New York, pp. 133–148.

Grossman, G. and Helpman, E. (1991) *Innovation and Growth in the Global Economy*. MIT Press, London.

Gusta, M., Smyth, S., Belcher, K., Phillips, P. and Castle, D. (2011) Economic benefits of genetically modified herbicide tolerant canola for producers. *AgBioForum* 14, 1–13.

Hackman, J. and Walton, R. (1986) Leading groups in organizations. In: Goodman, P. (ed.) *Designing Effective Work Groups*. Jossey-Bass, San Francisco, California, pp. 72–119.

Hall, P. (1999) *Cities in Civilization*. Orion, London.

Hamilton, A. (1790) *Report on Manufactures*.

Head, K., Ries, J. and Swenson, D. (1995) Agglomeration benefits and location choice: evidence from Japanese manufacturing investments in the United States. *Journal of International Economics* 38, 223–247.

Heckscher, E. (1919) The effect of foreign trade on the distribution of income. *Ekonomisk Tidskriff* XXI, 497–512. [Translated as Chapter 13 in American Economic Association (1949) *Readings in the Theory of International Trade*. Blakiston, Philadelphia, Pennsylvania, pp. 272–300.]

Held, J. (1996) Clusters as an economic development tool: beyond the pitfalls. *Economic Development Quarterly* 10, 249–261.

Helliwell, J. (1994) Empirical linkages between democracy and economic growth. *British Journal of Political Science* 24, 225–248.

Henderson, V., Kuncoro, A. and Turner, M. (1995) Industrial development in cities. *Journal of Political Economy* 103, 1067–1090.

Hersey, P. and Blanchard, K. (1977) *Management of Organizational Behavior: Utilizing Human Resources*, 3rd edn. Prentice Hall, Upper Saddle River, New Jersey.

Hetherington, K. (2007) *Capitalism's Eye: Cultural Spaces of the Commodity*. Routledge, New York.

Hicks, J. (1932) *The Theory of Wages*. Macmillan, London.

Holbrook, J. and Wolfe, D. (eds) (2000) *Innovation, Institutions and Territory: Regional Innovation Systems*. McGill-Queen's University Press, Kingston, Ontario, Canada.

Holbrook, J. and Wolfe, D. (eds) (2002) *Knowledge, Clusters and Regional Innovation: Economic Development in Canada*. McGill-Queen's University Press, Kingston, Ontario, Canada.

Hughes, T. (1989) *American Genesis: a Century of Invention and Technological Enthusiasm, 1870–1970*. Viking Penguin, New York.

Hunter, F. (1953) *Community Power Structure: a Study of Decision Makers*. University of North Carolina Press, Chapel Hill, North Carolina.

Irwin, D. (1996) *Against the Tide: an Intellectual History of Free Trade*. Princeton University Press, Princeton, New Jersey.

Isaksen, A. (1996) Towards increased regional specialization? The quantitative importance of new industrial spaces in Norway, 1970–1990. *Norwegian Geographical Journal* 50, 113–123.

Isaksen, A. (2004) Knowledge-based clusters and urban location: the clustering of software consultancy in Oslo. *Urban Studies* 41, 1157–1174.

Jacobs, J. (1969) *The Economy of Cities*. Random House, New York.

Jacobs, J. (1984) *Cities and the Wealth of Nations*. Vintage, New York.

Jensen, M., Johnson, B., Lorenz, E. and Lundvall, B.-Å. (2007) Forms of knowledge and modes of innovation. *Research Policy* 36, 680–693.

Johnson, D. (2002) *Address to the Virtual College of Biotechnology Mixer*, Saskatoon, 25 September.

Johnson, S. (2010) *Where Good Ideas Come From: the Natural History of Innovation*. Penguin Group, New York.

Karwandy, J. (2008) Perspectives on industrial clustering and the product, resource and knowledge based views of management. MSc Thesis, University of Saskatchewan, Saskatoon, Saskatchewan, Canada.

Keynes, J.M. (1936) *The General Theory of Employment, Interest and Money* <http://ambidextrouscivicdiscourse.com/wp-content/uploads/2010/10/The-General-Theory-of-Employment-Interest-and-Money.pdf>.

Klepper, S. (2004) *Agglomeration through Spinoffs: How Detroit Became the Capital of the US Automobile Industry*. Working Paper, Carnegie Mellon University, Pittsburgh, Pennsylvania.

Kline, S. and Rosenberg, N. (1986) An overview of innovation. In: Landau, R. and Rosenberg, N. (eds) *The Positive Sum Strategy: Harnessing Technology for Economic Growth*. National Academy Press, Washington, DC, pp. 275–306.

Knoke, D. (1999) Organizational Networks and Corporate Social Capital. In: Leenders, R. and Gabbay, S. (eds) *Corporate Social Capital and Liability*. Kluwer, Boston, Massachusetts.

Knoke, D. and Kuklinski, J. (1982) *Network Analysis*. Series: Quantitative Applications in the Social Sciences. Sage Publications, Thousand Oaks, California.

KPMG, DRI/McGraw-Hill and IMPAX Policy Services International (1996) *Building Technology Bridges: Cluster-based Economic Development for Western Canada*. Western Economic Diversification, Ottawa, Canada.

Krebs, V. (2002) Mapping networks of terrorist cells. *Connections* 24, 43–52.

Krebs, V. (2005) InFlow 3.1 Network Mapping Software <http://www.orgnet.com/inflow3.html>.

Krugman, P. (ed.) (1986) *Strategic Trade Policy and International Economics*. MIT Press, Cambridge, Massachusetts.

Krugman, P. (1991a) *Geography and Trade*. MIT Press, Cambridge, Massachusetts.

Krugman, P. (1991b) Increasing Returns and Economic Geography. *Journal of Political Economy* 99, 483–499.

Krugman, P.R. (1996) Making sense of the competitiveness debate. *Oxford Review of Economic Policy* 12, 17–25.

Krugman, P. (1998a) What's new about the new economic geography? *Oxford Review of Economic Policy* 14, 7–17.

Krugman, P. (1998b) *The Role of Geography in Development.* Paper prepared for the Annual World Bank Conference on Development Economics, Washington, District of Columbia, 20–21 April.

Krugman, P. and Venables, A. (1996) Integration, specialization, and adjustment. *European Economic Review* 40, 959–967.

Kuhn, T. (1970) *The Structure of Scientific Revolutions*, 2nd edn. University of Chicago Press, Chicago, Illinois.

Lagnevik, M., Sjöholm, I., Lareke, A. and Östberg, J. (2003) *The Dynamics of Innovation Clusters: a Study of the Food Industry.* Edward Elgar, Cheltenham, UK.

Landry, R. (2004) Local vs global links: What is the trade-off? Presentation at the *Innovation Systems Research Network Sixth Annual Meeting*, Vancouver, Canada <http://www.utoronto.ca/isrn/publications/NatMeeting/NatSlides/Nat04/Landry04_Links.ppt>.

Lawson, C. and Lorenz, E. (1999) Collective learning, tacit knowledge, and regional innovative capacity. *Regional Studies* 33, 305–317.

Leadbeater, C. (2005) *Charles Leadbeater on Innovation.* TED Talks. <http://www.ted.com/talks/charles_leadbeater_on_innovation.html>.

Leamer, E. and Storper, M. (2001) The economic geography of the Internet age. *Journal of International Business Studies* 32, 641–665.

Leydesdorff, L. and Etzkowitz, H. (1998) The Triple Helix as a model for innovation studies. *Science and Public Policy* 25, 195–203.

Leydesdorff, L. and Meyer, M. (2006) Triple Helix indicators of knowledge-based innovation systems: Introduction to the special issue. *Research Policy* 35, 1441–1449.

List, F. (1841/1909) *The National System of Political* Economy. Translated by S. Lloyd. Longmans, Green and Co., New York.

Lucas, R. (1988) On the mechanics of economic development. *Journal of Monetary Economics* 22, 30–42.

Lundvall, B.-Å. (1988) Innovation as an interactive process: from user–producer interaction to the national system of innovation. In: Dosi, G., Freeman, C., Silverberg, G. and Soete, L. (eds) *Technical Change and Economic Theory.* Pinter Publishers, London, pp. 349–369.

Lundvall, B.-Å. (ed.) (1992) *National Systems of Innovation: Towards a Theory of Innovation and Interactive Learning.* Pinter, London.

Lundvall, B.-Å. (2004) *Why the new economy is a learning economy.* DRUID Working Paper 04-01. Aalborg University, Aalborg, Denmark.

Lundvall, B.-Å. and Borrás, S. (1998) *The Globalising Learning Economy: Implications for Innovation Policy.* TSER Studies, DG XII. European Union, Luxembourg.

Lundvall, B.-Å., Lorenz, E. and Drejer, I. (2004) *How Europe's Economies Learn.* Report from the Loc Nis Policy Workshop. European Commission, Brussels.

Mahoney, J. (1992) The choice of organisational form: vertical financial ownership versus other methods of vertical integration. *Strategic Management Journal* 13, 559–584.

Maillat, D. (1988) La revitalisation des régions de tradition industrielle: le rôle des PME et du milieu. *Revue International PME* 1, 1.

Maillat, D. (1991) The innovation process and the role of the milieu. In: Bergman, E., Maier, G. and Tödling, F. (eds) *Regions Reconsidered: Economic Networks, Innovation and Local Development in Industrialized Countries*. Mansell Publishing, London/New York, pp. 103–118.

Maillat, D. (1995) Territorial dynamic, innovative milieus and regional policy. *Entrepreneurship and Regional Development* 7, 157–165.

Maine, E., Shapiro, D. and Vining, A. (2008) The role of clustering in the growth of new technology-based firms. *Small Business Economics* 34, 127–146.

Malanga, S. (2004) The curse of the creative class. *City Journal* Winter <http://www.city-journal.org/html/14_1_the_curse.html>.

Malecki, E. (1997) *Technology and Economic Development: the Dynamics of Local, Regional and National Competitiveness*. Longman, Toronto, Ontario, Canada.

Malmberg, A. and Maskell, P. (1997) Towards an explanation of regional specialization and industry agglomeration. *European Planning Studies* 5, 25–41.

Malmberg, A. and Maskell, P. (2002) The elusive concept of localization economies: towards a knowledge-based theory of spatial clustering. *Environment and Planning A* 34, 429–449.

Mankiw, N., Romer, D. and Weil, D. (1992) A contribution to the empirics of economic growth. *Quarterly Journal of Economics* CVII, 407–437.

Mann, R. (1959) A review of the relationship between personality and performance in small groups. *Psychological Bulletin* 66, 241–270.

Markusen, A. (1999) Fuzzy concepts, scanty evidence, policy distance: the case for rigour and policy relevance in critical regional studies. *Regional Studies*, 33, 869–884.

Markusen, A. (2006) Urban development and the politics of a creative class: evidence from the study of artists. *Environment and Planning* A 38, 1921–1940.

Marshall, A. (1890) *Principles of Economics*. Macmillan, London.

Marshall, A. (1923) *Money Credit and Commerce*. Macmillan, London.

Martin, R. and Sunley, P. (2003) Deconstructing clusters: chaotic concept or policy panacea? *Journal of Economic Geography* 3, 5–35.

Maskell, P. and Lorenzen, M. (2003) *The Cluster as Market Organization*. DRUID Working Paper no. 03-14 <http://ideas.repec.org/p/aal/abbswp/03-13.html>.

Maskell, P. and Kebir, L. (2005) *What Qualifies as a Cluster Theory?* DRUID working paper no. 05-09. <http://www.druid.dk/wp/pdf_files.org/05-09.pdf>.

Mead, S. (2001) Using social network analysis to visualize project teams. *Project Management Journal* 32, 32–38.

Metcalfe, J. (1995) Technology systems and technology policy in an evolutionary framework. *Cambridge Journal of Economics* 19, 25–46.

Mills, C.W. (1956) *The Power Elite*. Oxford University Press, New York.

Mitchell, T., Holtom, B., Lee, T., Sablynski, C. and Erez, M. (2001) Why people stay: using job embeddedness to predict voluntary turnover. *Academy of Management Journal* 44, 1102–1121.

Moore, G. and Davis, K. (2004) Learning the Silicon Valley way. In: Bresnahan, T. and Gambardella, A. (eds) *Building High-Tech Clusters: Silicon Valley and Beyond*. Cambridge University Press, Cambridge, pp. 7–39.

Mote, J. (2005) R&D ecology: using 2-mode network analysis to explore complexity in R&D environments. *Journal of Engineering & Technology Management* 22, 93–111.

Mowery, D. and Oxley, J. (1995) Inward technology transfer and competitiveness: the role of national innovation systems. *Cambridge Journal of Economics* 19, 67–93.

Munn-Venn, T. and Voyer, R. (2004) *Clusters of Opportunity, Clusters of Risk.* Conference Board of Canada, Ottawa, Canada <http://www.conferenceboard.ca/e-Library/abstract.aspx?did=768>.

National Research Council Canada (1997) *Bi-annual Report.* National Research Council, Ontario, Canada.

National Research Council Canada (2004) *The NRC Model of Cluster Development* <http://www.nrc-cnrc.gc.ca/doingbusiness/clusters_2> (archived report).

National Research Council Canada (2005) *Building Technology Clusters Across Canada.* National Research Council, Ontario, Canada <http://www.nrc-cnrc.gc.ca/eng/news/nrc/2005/05/05/technology-clusters.html>.

Nelson, R. (1993) *National Innovation Systems: a Comparative Analysis.* Oxford University Press, Oxford.

Nelson, R. and Winter, S. (1977) In search of useful theory of innovation. *Research Policy* 6, 36–76.

Niosi, J. and Bas, T. (2001) The competence of regions: Canada's clusters in biotechnology. *Small Business Economics* 17, 31–42.

Nonaka, I. and Takeuchi, H. (1995) *The Knowledge Creating Company: How Japanese Companies Create the Dynamics of Innovation.* Oxford University Press, New York.

Nordicity Group (1996) *Regional/Local Industrial Clustering: Theory and Lessons from Abroad.* National Research Council, Government of Canada.

North, D. (2005) *Understanding the Process of Economic Change.* Princeton University Press, Princeton, New Jersey.

Ohlin, B. (1933) *Interregional and International Trade.* Harvard University Press, Cambridge, Massachusetts.

Olson, M. (1965) *The Logic of Collective Action: Public Goods and the Theory of Groups.* Harvard University Press, Cambridge, Massachusetts.

OECD (1993) *Frascati Manual*, 3rd edn. Organisation for Economic Co-operation and Development, Paris.

OECD (1996) *The Knowledge-based Economy.* Organisation for Economic Co-operation and Development, Paris <http://www.oecd.org/dataoecd/51/8/1913021.pdf>.

OECD (1997) *National Innovation Systems.* Organisation for Economic Co-operation and Development, Paris <http://www.oecd.org/dataoecd/35/56/2101733.pdf>.

OECD (2005) *The Oslo Manual: Guidelines for Collecting and Interpreting Innovation Data*, 3rd edn. Organisation for Economic Co-operation and Development, Paris <http://www.oecdbookshop.org/oecd/display.asp?K=5LGPBVQFQ4G5&lang=EN&sort=sort_date%2Fd&sf1=Title&st1=oslo+manual&sf3=SubjectCode&st4=not+E4+or+E5+or+P5&sf4=SubVersionCode&ds=oslo+manual%3B+All+Subjects%3B+&m=3&dc=7&plang=en.

OECD (2010) *OECD Innovation Strategy: a Head Start on Tomorrow.* Organisation for Economic Co-operation and Development, Paris.

Ottaviano, G. and Thisse, J.-F. (2003) Agglomeration and economic geography. In: Henderson, J. and Thisse, J.-F. (eds) *Handbook of Urban and Regional Economics.* North Holland, New York.

Ouimet M., Landry, R. and Amara, N. (2007) Network positions and efforts to innovate in a small Canadian optics and photonics cluster. *International Journal of Entrepreneurship and Innovation Management* 7, 251–271.

Padmore, T. and Gibson, H. (1998) Modelling systems of innovation: II. A framework for industrial cluster analysis in regions. *Research Policy* 26, 625–641.

Patel, P. and Pavitt, K. (1994) The nature and economic importance of national innovation systems. *STI Review* 3, 9–32.

Patrucco, P. (2005) The emergence of technology systems: knowledge production and distribution in the case of Emilian plastics district. *Cambridge Journal of Economics* 29, 37–56.

Penrose, E. (1959) *The Theory of the Growth of the Firm.* Oxford University Press, Oxford.

Phillips, P. (2000) Intellectual property rights, canola and public research in Canada. In: Santaniello, V., Evenson, R., Zilberman, D. and Carlson, G. (eds) *Agriculture and Intellectual Property Rights.* CAB International, Wallingford, UK.

Phillips, P. (2001) The impact of location on production. In: Phillips, P.W.B. and Khachatourians, G.G. (eds) *The Biotechnology Revolution in Agriculture: Invention, Innovation and Investment in the Canola Sector.* CAB International, Wallingford, UK, pp. 161–186.

Phillips, P. (2002) Regional systems of innovation as a modern R&D entrepôt: the case of the Saskatoon biotechnology cluster. In: Chrisman, J. and Holbrooke, A. (eds) *Innovation, Entrepreneurship, Family Business, and Economic Development: a Western Canadian Perspective.* University of Calgary Press, Calgary, Canada, pp. 31–58.

Phillips, P. (2007) *Governing Transformative Technological Innovation: Who's in Charge?* Edward Elgar, Oxford, UK.

Phillips, P. and Khachatourians, G. (2001) *The Biotechnology Revolution in Global Agriculture: Invention, Innovation and Investment in the Canola Sector.* CAB International, Wallingford, UK.

Phillips, P. and Webb, G. (2008) Talent, tolerance and community in Saskatoon. *Proceedings of the ISRN National Conference*, Montreal, Canada, 5 May.

Phillips, P. and Webb, G. (2013a) Cooperative and partnership governance in Saskatoon. In: Bradford, N. (ed.) *Civic Engagement and Collaborative Governance in Canadian City-Regions.* ISRN Series, Vol. III. University of Toronto Press, Toronto, Canada (in press).

Phillips, P. and Webb, G. (2013b) Talent, tolerance and community in Saskatoon. In: Grant, J.L. (ed.) *Attracting and Retaining Talented and Creative Workers in Canadian Cities: The Social Dynamics of Economic Innovation.* ISRN Series, Vol. II. University of Toronto Press, Toronto, Canada (in press).

Phillips, P. and Webb, G. (2013c) Social dynamics, diversity and physical infrastructure in creative, innovative communities: the Saskatoon case. In: Wolfe, D. (ed.) *Innovation and Knowledge Flows in Canadian Cities.* ISRN Series Vol. I. University of Toronto Press, Toronto, Canada (in press).

Phillips, P. and Webb, G. (2013d) Saskatoon: from small town to global hub. In: Wolfe, D. and Gertler, M. (eds) *Innovation and Knowledge Flows in Canadian Cities.* ISRN Series, Vol. IV. University of Toronto Press, Toronto, Canada (in press).

Phillips, P., Parchewski, J., Procyshyn, T., Ryan, C., Karwandy, J. and Kihlberg, J. (2004) *Agricultural and Life-science Clusters in Canada: An Empirical and Policy Analysis*. Final project report for AAFC Study of the Saskatoon Cluster, 30 March <http://www.agbio-management.org>.

Phillips, P., Ryan, C., Karwandy, C., Procyshyn, T. and Parchewski, J. (2008a) The Saskatoon agricultural biotechnology cluster. In: Karlsson, C. (ed.) *Handbook of Research on Clusters: Theories, Policies and Case Studies*. Edward Elgar, Cheltenham, UK.

Phillips, P., Boland, W. and Jiang, B. (2008b) *Public–Private Partnerships in the Global Agri-food System: an Emerging Australian Model?* Working paper in Governing Transformative Innovation <http://www.usask.ca/politic/phillips/Governance.htm>.

Phillips, P., Boland, W. and Ryan, C. (2009) *Centreless Governance for the Management of a Global R&D Process: Public–Private Partnerships and Plant-Genetic Resource Management*. Working paper 1-1-2009, OpenSIUC, Southern Illinois University, Carbondale <http://opensiuc.lib.siu.edu/cgi/viewcontent.cgi?article=1013&context=pn_wp>.

Phillips, P., Webb, G., Boland, B. and Ryan, C. (2010) *Research and Innovation Delivery Models in Agri-based Innovation Systems*. Report submitted to Innovation Policy Analysis, Research Branch, Agriculture and Agri-Food Canada.

Picciotto, R. (1997) Putting institutional economics to work: from participation to governance. In: Clague, C. (ed.) *Institutions and Economic Development: Growth and Governance in Less Developed and Post-Socialist Countries*. Johns Hopkins University Press, Baltimore, Maryland, pp. 343–367.

Porter, M.E. (1985) *Competitive Advantage: Creating and Sustaining Superior Performance*. Free Press, New York.

Porter, M. (1990) *The Competitive Advantage of Nations*. Free Press, New York.

Porter, M. (1998) *On Competition*. Harvard Business School Press, Boston, Massachusetts.

Powell, W. (1990) Neither market nor hierarchy: network forms of organization. In: Straw, B. and Cummings, L. (eds) *Research in Organizational Behaviour*. JAI Press, Greenwich, Connecticut, pp. 295–336.

Procyshyn, T. (2004) Saskatoon's agricultural biotechnology clusters and the Canadian Light Source: an assessment of the potential for cluster extension. MSc thesis, University of Saskatchewan, Saskatoon, Saskatchewan, Canada <http://www.usask.ca/politic/phillips/pdfs/Procyshyn%20-%20Thesis.pdf>.

Procyshyn, T. and Ryan, C. (2003) Hard measures and soft issues: a potential model for incorporating metrics into cluster based analysis. In: *Proceedings of the ISRN Annual Meeting*, Ottawa.

Putnam, R. (2000) *Bowling Alone: the Collapse and Revival of American Community*. Simon & Schuster, London.

Roelandt, T. and den Hertog, P. (eds) (1998) *Cluster Analysis and Cluster Based Policy in OECD Countries: Various Approaches, Early Results and Policy Implications*. OECD, The Hague/Utrecht, May. <http://www.innovation.lv/ris/latv/Bibl/OECD_FG_CLUST.pdf>.

Romer, P. (1990) Endogenous technological change. *Journal of Political Economy* 98, S71–S102.

Romer, P. (1994) The origins of endogenous growth. *Journal of Economic Perspectives* 8, 3–22.

Rosenfeld, S.A. (1995) *Industrial Strength Strategies: Regional Business Clusters and Public Policy*. Aspen Institute <http://www.aspeninstitute.org/publications?program=276&page=no>.

Rugman, A. and D'Cruz, J. (2000) *Multinationals as Flagship Firms: Regional Business Networks*. Oxford University Press, Oxford.

Rugman, A. and Verbeke, A. (2001) Subsidiary-specific advantages in multinational enterprises. *Strategic Management Journal* 22, 237–250.

Rugman, A. and Verbeke, A. (2002) Edith Penrose's contribution to the resource-based view of strategic management. *Strategic Management Journal* 23, 769–780.

Rullani, E. (2002) The industrial district (ID) as a cognitive system. In: Curzio, Q. and Fortis, M. (eds) *Complexity and Industrial Districts*. Springer Verlag, Berlin, Germany, pp. 35–61.

Ruttan, V. (2003) *Social Science Knowledge and Economic Development: an Institutional Design Perspective*. University of Michigan Press, Ann Arbor, Michigan.

Ryan, C.D. (2004) *Clusters and Complexity: Towards a New Epistemology for Knowledge Management and Innovation in the Saskatoon Agricultural Biotechnology Cluster*. Working paper on Governing Transformative Innovation <http://www.usask.ca/politic/phillips/pdfs/Comprehensive%20Paper%20-%20C.D.%20Ryan%20-%20June%202004.pdf>.

Ryan, C. (2008) *Evaluating Performance of Research Networks: a Socioeconomic Framework for Assessing Funded Research Projects*. VDM Verlag, Germany.

Ryan, C. (2011) *Evaluating Scholarly Networks for the Canadian Light Source Synchrotron: Knowledge, Entrepreneurial and People Advantage*. Report for the Canadian Light Source Synchrotron, 15 September.

Ryan, C. and Phillips, P. (2002) Industrial innovation and regional competitiveness in the agricultural biotechnology sector: a comparative analysis of innovation structures in North America, Europe and Australia. In: *5th International Conference on Chain and Network Management in Agribusiness and the Food Industry, Poster Session*, Noordwijk, The Netherlands, 5–7 June <http://www.usask.ca/politic/phillips/pdfs/Analysis%20Innovation%202002.pdf>.

Ryan, C. and Phillips, P. (2003) Intellectual property management in clusters: a framework for analysis? In: Wolfe, D. (ed.) *Clusters Old and New: the Transition to a Knowledge Economy in Canada's Regions*. McGill-Queen's University Press for the School of Policy Studies, Queen's University, Kingston and Montreal, Canada.

Ryan, C. and Phillips, P. (2004) Knowledge management in advanced technology industries: an examination of international agricultural biotechnology clusters. *Environment and Planning, Government, and Policy* 22, 217–232.

Ryan, C. and Smyth, S. (2010) Facilitating innovation in agricultural biotechnology: an examination of the Ag-West Biotech Governance Model, 1989–2004. *AgBioForum* 13, 183–193.

Ryan, C., Phillips, P. and Boland, W. (2011) Public–private partnerships and public policy: a comparative social network and public policy analysis on competing national systems of innovation. In: *Proceedings of the Triple Helix 9 Conference,* Stanford, California, 11–14 July.

Ryan, C., St Louis, M. and Phillips, P. (2012) *Incorporating Network Analysis into Evaluation of Innovation in 'Big Science': Networks of Scholars Affiliated with the Canadian Light Source Synchrotron*. Presentation at the ISPIM Conference. Barcelona, Spain, June.

Saxenian, A. (1990) Regional networks and the resurgence of Silicon Valley California. *Management Review* 33, 89–112.

Saxenian, A. (1994) *Regional Advantage: Culture and Competition in Silicon Valley and Route 128*. Harvard University Press, Cambridge, Massachusetts.

Schumpeter, J. (1934) *The Theory of Economic Development: an Inquiry into Profits, Capital, Credit, Interest, and the Business Cycle*. Harvard University Press, Cambridge, Massachusetts.

Schumpeter, J. (1939) *Business Cycles: a Theoretical, Historical and Statistical Analysis of the Capitalist Process*. McGraw-Hill, New York.

Schumpeter, J. (1954) *Capitalism, Socialism, and Democracy*. George Allen and Unwin, London.

Scott, J. (2001) Studying power. In: Nash, K. and Scott, A. (eds) *The Blackwell Companion to Political Sociology*. Blackwell Publishers, London.

Scott, W. (1995) *Institutions and Organisations*. Sage Publications, London.

Scottish Enterprise Network (2000) *The Clusters Approach* <http://www.scottish-enterprise.com/publications//shareddocs/social_economy_and_clusters.pdf> (archived report).

Seabright, P. (2004) *The Company of Strangers: a Natural History of Economic Life*. Princeton University Press, Princeton, New Jersey.

Sen, G. (1987) *Men and Things*. Inaugural Lecture delivered at the London School of Economics, 7 March. <http://www.compilerpress.ca/Competitiveness/Anno/Anno%20Sen%20Men%20and%20Things.htm>.

Simon, H. (1955) A behavioral model of rational choice. *Quarterly Journal of Economics* 69, 99–188.

Smith, A. (1776) *An Inquiry into the Nature and Causes of the Wealth of Nations*. Methuen and Co., London.

Solow, R. (1956) A contribution to the theory of economic growth, *Quarterly Journal of Economics* 70, 65–94.

Sölvell, O., Lindqvist, G. and Ketels, C. (2003) *The Cluster Initiative Greenbook*. The Competitiveness Institute, September <http://www.cluster-research.org/greenbook.htm>.

Spencer, G. and Vinodrai, T. (2006) *Innovation Systems Research Network City-Region Profile*. Program on Globalization and Regional Innovation Systems (PROGRIS), University of Toronto.

SREDA (1998) Biotechnology industry profiles. Mimeograph. Saskatchewan Regional Economic Development Authority, Saskatchewan, Canada.

Stabler, J. and Olfert, M.R. (1996) *The Changing Role of Rural Communities in an Urbanizing World: Saskatchewan – An Update to 1995*. Canadian Plains Research Centre, University of Regina, Saskatchewan, Canada.

Stern, S., Porter, M. and Furman, M. (2001) *The Determinants of National Innovative Capacity*. Working Paper no. 7876. National Bureau of Economic Research, Cambridge, Massachusetts <http://www.nber.org/papers/w7876>.

Stogdill, R. (1948) Personal factors associated with leadership: a survey of the literature. *Journal of Psychology* 25, 35–71.

Stone, D. (1989) Causal stories and the formation of policy agendas. *Political Science Quarterly* 104, 281–300.

Storper, M. (1997) *The Regional World: Territorial Development in a Global Economy*. Guilford Press, New York.

Storper, M. and Venables, A. (2004) Buzz: face-to-face contact and the urban economy. *Journal of Economic Geography* 4, 351–370.

Strange, S. (1988) *States and Markets*. Pinter, London.

Theodorakoupoulou, I. and Kalaitzandonakes, N. (1999) Structure and performance of private–public knowledge networks in plant biotechnology. In: Wolfe, S. and Zilberman, D. (eds) *Knowledge Generation and Technical Change: Institutional Innovation in Agriculture*. Kluwer-Plenum Academic Publishers, New York, pp. 177–192.

Valente, T.W. (1995) *Network Models of the Diffusion of Innovations*. Hampton Press, Cresskill, New Jersey.

Valente, T., Chou, C. and Pentz, M. (2007) Community coalition network as systems: effects of network change on adoption of evidence based prevention. *American Journal of Public Health* 97, 880–886.

van der Linde, C. (2002) *Findings from the Cluster Meta-study*. Institute for Strategy and Competitiveness, Harvard Business School, Boston, Massachusetts <http://www.isc.hbs.edu/MetaStudy2002Prz.pdf>.

Venables, A. (1996) Localization of industry and trade performance. *Oxford Review of Economic Policy* 12, 52–60.

Venables, A. (2006) Shifts in economic geography and their causes. *Economic Review, Federal Reserve Bank of Kansas City* Q IV, 61–85.

Wasserman, S. and Faust, K. (1994) *Social Network Analysis*. Cambridge University Press, New York.

Webb, G. (2009) Creative social entrepreneurs, social capital and collaborative governance: a Saskatoon based analysis. MSc Thesis, University of Saskatchewan <http://library.usask.ca/theses/available/etd-07152009-105147/unrestricted/GraemeWebbThesis.pdf>.

Williamson, O. (1981) The economics of organization: transaction cost approach. *American Journal of Sociology* 87, 548–577.

Williamson, O. (1985) Reflections of the new institutional economics. *Journal of Institutional and Theoretical Economics* 141, 187–195.

Wolfe, D. (2002) Innovation policy for the knowledge-based economy: from the Red Book to the White Paper. In: Doern, G. (ed.) *How Ottawa Spends, 2001–2002*. Oxford University Press, Toronto, Canada.

Wolfe, D. (2003) *Clusters Old and New: the Transition to a Knowledge Economy in Canada's Regions*. McGill-Queen's University Press for the School of Policy Studies, Queen's University, Kingston, Ontario, Canada.

Wolfe, D. and Gertler, M. (2004) Clusters from the inside and out: local dynamics and global linkages. *Urban Studies* 41, 1055–1077.

Wolfe, D. and Lucas, M. (eds) (2004) *Clusters in a Cold Climate: Innovation Dynamics in Canada's Diverse Economy*. School of Policy Studies, Queen's University and McGill-Queen's University Press, Montreal and Kingston, Canada.

Wolfe, D., Davis, C. and Lucas, M. (2004) Global networks and local linkages: an introduction. In: Wolfe, D. and Lucas, M. (eds) *Global Networks and Local Linkages: the Paradox of Cluster Development in an Open Economy*. McGill-Queen's University Press, Montreal and Kingston, Canada, pp. 1–24.

World Bank (2008) *World Development Report 2009: Reshaping Economic Geography*. World Bank, Washington, DC <http://econ.worldbank.org/WBSITE/EXTERNAL/EXTDEC/EXTRESEARCH/EXTWDRS/0,,contentMDK:20227703~pagePK:478093~piPK:477627~theSitePK:477624,00.html>.

Zaccaro, S. (2007) Trait-based perspectives of leadership. *American Psychologist,* 62, 6–16.

Zeller, C. (2001) Clustering in biotech: a recipe for success? Spatial patterns for growth of biotechnology in Munich, Rhineland, and Hamburg. *Small Business Economics* 17, 123–141.

Zilberman, D., Yarkin, C. and Heiman, A. (1997) Agricultural biotechnology: economic and international implications. In: *Proceedings of the XXIII International Conference of Agricultural Economists, Food Security, Diversification, and Resource Management: Refocusing the Role of Agriculture*, Sacramento, California, 10–16 August.

Zucker, L., Darby, M. and Brewer, M. (1998) Intellectual human capital and the birth of U.S. biotechnology enterprises. *American Journal of Economics* 88, 290–306.

Index

Page numbers in **bold** refer to figures and tables.